THE MASTER
HANDBOOK OF
ELECTRICAL WIRING

Other TAB books by the author

No. 1019
$10.95

THE MASTER HANDBOOK OF ELECTRICAL WIRING
BY ART MARGOLIS

TAB BOOKS
BLUE RIDGE SUMMIT, PA. 17214

FIRST EDITION

FIRST PRINTING—AUGUST 1978

Copyright © 1978 by TAB BOOKS

Printed in the United States
of America

Library of Congress Cataloging in Publication Data

Margolis, Art.
   The master handbook of electrical wiring.

   Includes index.
   1. Electric wiring—Handbooks, manuals, etc. I. Title.
TK3201.M35   621.319'2   78-13609
ISBN 0-8306-9889-2
ISBN 0-8306-1019-7 pbk.

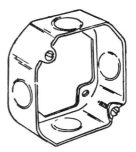

Preface

Some mechanically inclined people are disposed to view the subject of electrical wiring as a simple one. The notion is far from true. Although a certain amount of brawn is often involved in a wiring project, the elements of skill, thought and care are required in far greater degree. A safe and proper electrical installation demands a level of expertise that none can deprecate.

The material within these covers is designed to meet the desires for information and instruction of a multiple audience. The book addresses both the non-professional and the professional Some readers will be interested in pursuing the career of an electrician, in qualifying for Apprentice, Journeyman or Master status, in earning a living through the installation of electrical systems. Furthermore, the material will prove extremely useful to the electronic technician and add a full new dimension to his repertoire.

The preponderance of the readership, however, is likely to consist of do-it-yourselfers and amateur electricians. Some members of this category will be novices who wish to learn how to add a switch, receptacle or simple circuit to their households. Others among the group will be looking for guidance in planning—even hooking up (in accordance with official inspection)—the entire electrical system in a new home or in revamping an outmoded system in an old home.

Whenever instruction, advice and recommendations are directed to the professional, the author intends for the non-professional to attend the message closely. As in all do-it-yourself activity the amateur becomes proficient by study and imitation of the professional. The more the amateur learns about the professional electrician's field the better he will be equipped to carry out his own projects.

This book has been written simply but is complete enough to be of considerable value to each segment of its readership. It discusses tools, elementary theory and hardware in Chapters 1 through 8. Chapters 9 through 12 delve into circuit considerations. Chapters 13 through 17 put to work all the material in the first 12 chapters: residential, farming and commercial wiring jobs are calculated and the installations discussed.

The 1978 National Electrical Code is being printed as this book is being completed. The material here has been keyed to conform with the new Code as closely as possible. If you read these pages from beginning to end, you will have covered the same general areas that are taught in vocational schools and professional-type classes throughout the country.

Art Margolis

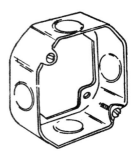

Contents

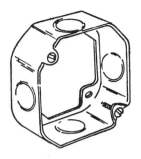

INTRODUCTION

The wireman who installs and repairs electrical systems in residences, farms, shops and factories deals with the last link in a long chain. The electric utility company delivers three or four wires to him. The wires are loaded with electrical energy. The wireman adapts these wires to supply the electrical system he has installed or repaired.

The Overall Electric System

This book discusses the electrical wiring systems (Fig. 1) the wireman works with. These systems begin with what is called the service. The service is attached to the wires of the utility company. The service is the last link between the utility company and the realm of the wireman.

Before beginning a study of the electrical systems needed in home and business, it is desirable to have a picture in your mind of the general electrical network from the utility company's generating plant to the home or business service. This introduction presents an overall view of the major components in the network, showing how electricity is generated, transmitted, changed around for various uses, delivered to different distribution systems and conveyed to the service the wireman installs.

BASIC PARTS OF THE SYSTEM

The typical energy producing and delivery system is composed of the following major parts. First of all there are the generating

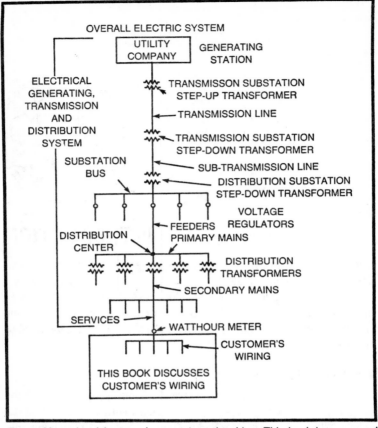

Fig. 1. Most electricians work on customer's wiring. This book is concerned mostly with this area.

stations. These are the gigantic, space age appearing installations throughout the nation. They contain engine or turbine generators that are powered by steam. The steam is produced in furnaces fueled by oil, coal, natural gas or atomic energy.

Once the turbines generate electricity, the voltage is attached to transmission lines. The transmission lines connect to substations. Out of the substations come feeder lines containing the electrical energy. The feeders go to what are called the primary mains. At the end of the mains are distribution transformers. Secondary mains emanate from the transformers. The secondary mains run up and down the streets and alleys near homes and business places. These mains are tapped for the service to individual buildings.

This is a general description. The actual electrical networks can be quite complex. There can be a number of generating stations and tens of substations. The system can be large or small, depending on the number of customers.

ELECTRIC GENERATORS

The generators at the utility company's plant are able to transform the energy of coal, oil, atom or moving water into electricity. For the generator to produce electricity, its turbine or water wheel must be driven. (The steam engine generators of past years have now all but disappeared from the scene.)

The modern turbine generator is usually driven by steam. The steam is produced from heat. The heat is created by burning coal, oil or gas, or by harnessing the energy released in atomic fission. Once the steam is made, it is used to turn the blades of the turbine.

Most steam plants have high smoke stacks which discharge sulphur dioxide and particulate wastes into the atmosphere and are a source of pollution. The nuclear power plant is an exception, since it produces no fossil fuel residues. However, nuclear plants present other hazards, such as radiation. The generation of electrical energy has its problems and the conversion of fuel to electricity is a major international worry.

About one fifth of the nation's electric generators are hydroelectric. They are located at waterfalls. The descending stream of water turns water wheels of sophisticated design and the need for fuel is eliminated. The hydroelectric plant has a lot of advantages which are the result of its not needing fuel. However, dams must be built and land flooded and the overall expense is high. More importantly, there are not enough suitable sites to generate the power that the consumers demand.

Besides the large generator stations, there are many small ones. For example, a large industrial plant might find it feasible to build its own on-site electric generator. The generator would make the plant independent of the utility company.

Although most smaller plants use fossil fuel or water wheels to generate electricity, some use internal combustion engines. These engines are usually powered by diesel fuel, but sometimes by gasoline.

Most generating plants produce three-phase alternating current at a precise frequency of 60 hertz. Sixty hertz is easy to transmit over the extensive distances between plant and users. In foreign countries nearly all electric power is produced at 50 hertz. This presents some conversion problems on occasion.

The only exception to 60 hertz production in this country is the output of hydroelectric plants. Often these plants are located at great distances from the consumer, the reason being that the plants must be built by the side of a waterfall. The city where the power will be consumed might be 100 miles or more from the plant. A large voltage drop loss could occur between plant and city, even at the low 60 hertz frequency. To prevent this, the transmission frequency would be lowered well below 60 hertz and then transformed back to 60 hertz at the city.

TRANSMISSION SYSTEM

The voltage produced at the generator might be of the order of 13,800. This seems like a high voltage, but if it were transmitted over the wires for any significant distance, a large voltage drop would take place, resulting in a great loss of power.

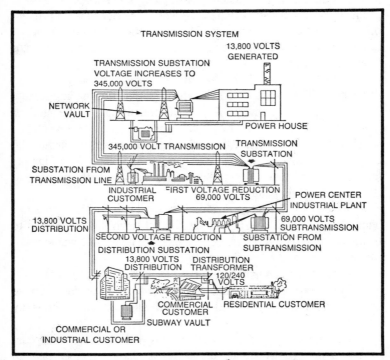

Fig. 2. The typical electrical transmission system has a number of step up and step down transformer substations between the power house and the customer.

Therefore, these 13,800 volts are passed through a transmission transformer before heading out across the transmission lines. The transformer substation steps up the 13,800 volts to 345,000. Then the voltage is sent out over the transmission lines.

The transmission lines are copper conductors and they are strung from pole to pole across the open spaces of our country. The poles and lofty towers are a familiar sight throughout the nation. The conductors are pressured with 345,000 volts. If you look up at the poles, you'll observe strings of porcelain disk insulators. The insulators hold the lines away from the pole's superstructure. At the top of the superstructure small static wires are attached. These wires are bare and connect from the metal on the pole to ground. They are lightning shields and arrestors.

Large cities do not permit the high transmission voltage to be strung overhead. There is too much of a possibility of wire breakage. High voltage lines falling onto city streets could cause great damage and death. All of the high voltage lines in cities must be run underground. The lines pass through underground tunnels. The lines look like pipes. They are actually pipes containing three well insulated and shielded cables. They transport large quantities of three-phase power throughout the city (Fig. 2).

SUBSTATIONS IN GENERAL

A substation is essentially a large transformer system. The transmission lines, whether overhead or underground, terminate at substations. Some large industrial customers of a utility might have their own substations. These customers will receive the high voltage and use the substation to convert the voltage for their particular needs. Normally, though, it is the utility company's substation which does the transforming, whether the customer be industrial or residential.

A substation's purpose is, first of all, to step down the high voltage in accordance with the needs of the consumer. For instance, the 345,000 volts can be stepped down to 69,000 volts at the transmission substation; then the voltage can be sent to small industrial plants which can utilize the 69,000 volts.

A second substation can step down the remainder of the voltage from 69,000 to 13,800. The voltage can then be distributed to an even greater number of customers who are able to utilize the 13,800 volts. A third transformer can step down the voltage even more, all the way down to 240 volts to suit the needs of a residence.

In addition, substations can have the capability of changing the frequency. For example, the voltage as mentioned could be transmitted at a lower frequency than 60 hertz. The substation could have the capability of changing that lower frequency back to the desired 60 hertz. There are other functions a substation can perform, including switching the circuits on and off.

SUBSTATION TYPES

There are three general types of substation, as sketchily described in the last section. There are transmission substations, subtransmission substations, and distribution substations.

The substations are differentiated by the function of their transformers. The most important of their functions is to change the generated voltage from one level to another. The transformer is able to either step up or step down the voltage, according to its design.

The transmission substation is a step up voltage type. It is located near the generator. The generator in a large plant typically produces 13,800 volts. The transmission substation steps up the 13,800 to the six-figure range, usually between 115,000 and 765,000 volts. The higher the voltage the easier it is transmitted over long distances.

The subtransmission substation is located somewhere along the route of the transmission. It is a step down station. It steps down the voltage to an intermediate voltage. Typical voltage levels at this stage are 23,000, 34,500 and 69,000 volts. These intermediate substations act as a distribution source for the local distribution substations.

Distribution substations are located near the consumers. These substations step down the voltage even further. The output of these substations is usually 4,160 volts. This voltage is distributed to the street wiring. It is the voltage that travels to the pots on the poles behind your house or beneath the ground to the concrete pads holding the small backyard distributor transformers. (Street Wiring is discussed in Chapter 18.) The pots and backyard transformers convert the 4,160 volts to the 240/120 volts needed in homes.

SUBSTATION TRANSFORMERS

The substation transformers are power transformers. They can handle prescribed amounts of energy. The energy is measured in kilovolt amps called kVA. This measurement resembles kilowattage, called kW. Both kVA and kW are measures of electrical energy. The transformer is rated in kVA. Typical power transformers in a transmission substation range between 50,000 kVA and 1,000,000 kVA. They handle the high transmission voltages.

The energy generated is three-phase. This is really three separate voltages traveling along the same wires. However, one transformer winding cannot handle the three phases effectively. Three transformer windings are needed. One for each phase.

Most substations use three-phase transformers. A three-phase transformer has separate windings for each phase. They are contained in the same transformer housing, each winding with its own terminal. It is not unlike three separate transformers.

Some substations are built with three single-phase transformers, one for each phase. The three transformers are connected together into a three-phase bank. The operation of three single-phase transformers is almost identical to the operation of the three-phase transformer. Often a spare transformer is installed in the substation, too. If the single-phase transformers fail, the service can be restored quickly.

SUBSTATION PROTECTION

If a substation fails, a lot of customers lose service. Substations are equipped with a number of protective devices: lightning arrestors, circuit breakers, fuses, disconnect switches, shunt reactors, and special capacitors. In addition, there are protection relays, and batteries to supply some power during emergencies.

LIGHTNING ARRESTORS

Lightning arrestors are designed to protect the substation and surrounding lines from lightning strikes. They also provide protection from the sudden voltage surges that occur during switching activity.

The lightning arrestors are installed close to the power transformers, since they are a vitally important part of the substation. They can be seen atop the transformers on the transmission line terminals.

The lightning arrestors are tall slim devices with a high-voltage look. They contain semiconductor materials that allow the current from the lightning strike to pass harmlessly to ground. The current cannot rebound out of ground because of the properties of the semiconductor materials.

CIRCUIT BREAKERS

The circuit breakers in the substation are large devices. They are designed to handle large amounts of current and high voltages. Some circuit breakers stand much taller than a man. The breaker mechanism is encased in a steel box. The box can be filled with oil, air, or gas, or it can contain a vacuum.

The breakers switch electric circuits and equipment in and out of the system. The contacts of the breaker are opened and closed remotely by linkages made of insulative materials. The contacts are actuated by springs, magnets or compressed air.

When trouble strikes, the breakers open automatically. Compressed air is usually the activating force. When the breakers have to be operated manually, a remote control switch with its own electricity opens or closes the large contacts.

FUSES

High voltage fuses can also be used to protect the electrical system. Usually, the fuses will blow if a fault develops in a substation transformer. The fuse protects the system from the transformer.

In the three-phase system there are three power fuses near the disconnect switch. The disconnect switch operates all three phases.

DISCONNECT SWITCHES

The substation must have switches so that the station can be shut down during emergencies or repair work. The control devices are called disconnect switches and are designed to turn off or on the high voltage that arrives at the substation from the transmission lines. The substation's output to the consumers is not usually switched. If it is, specially constructed switches are needed. It is easier to use a disconnecting device on the transmission line side of the station.

The switches are large. Some types are operated by turning a geared-down crank handle. Turning the handle moves rods slowly through the gear arrangement and rotates the insulators. This causes the switch blades to separate gradually and opens the substation input circuit.

Both the disconnect switches and the circuit breakers are able to open the substation electric circuit. If the circuit breaker should open for any reason, the disconnect switch can also be opened as an extra safety measure. The disconnect switch is important because its open or closed condition is visible. On the contrary, the state of the circuit breaker contacts cannot be seen.

Once the circuit breaker has been safely reset, the disconnect switch can be cranked closed. The disconnect switch is designed to isolate the substation from the transmission current.

SHUNT REACTORS

Shunt reactors are installed in many substations. They are large transformers with a single winding for all three phases. They are connected to all three phases. They serve as an inductive reactance.

As the transmission lines carry high voltage from generator station to substations, a capacitive reactance builds up between the long lines and earth. This reactance can become extremely high and cause great loss to the utility company.

To neutralize the capacitive reactance, large inductive reactances are introduced at the substation. The shunt reactors are connected to the transmission through a disconnect switch.

SUBSTATION CAPACITORS

Capacitors are installed in substations. While the transmission lines produce capacitive reactance, the customer load side produces inductive reactance. These inductive reactances can cause large voltage drop losses to occur. These losses are absorbed by the utility company, not by the consumer. Understandably, the utility company takes measures to eliminate the losses.

The company installs banks of capacitors. The capacitors sit in racks and are connected to all three phases and ground. They are designed to neutralize the inductance loads. The capacitors are installed in metal cabinets and each is fitted with a disconnect switch.

PROTECTIVE RELAYS

Protective relays are installed in the substations. They are found in the control panel. Also in the control panel are meters, switches, indicating lights, and other types of control devices.

The relays do two jobs. One, they operate to identify sources of trouble. Two, they remove the equipment they are attached to from the electrical system.

The relays are used in transmission as well as distribution circuits. The relays are carefully designed and arranged to perform all types of jobs as required by the electrical system to which they are attached.

BATTERIES

There are racks of batteries in substations. Perhaps you'll wonder why batteries are needed since the substation is supplied with all of that electricity traveling through the transmission lines.

Batteries supply direct current (DC) voltages. The DC voltages are needed to operate circuit breakers and other pieces of equipment. The direct current can be made from the incoming alternating current (AC), but in many cases it is more economical to utilize the storage battery.

During system shutdown or emergency, the alternating current is stopped before it gets into the substation. The storage battery still supplies the circuit breakers and other equipment during the substation shutdown. The batteries are unaffected by the shutdown.

Batteries also supply limited amounts of emergency power. The batteries cannot supply the consumer loads, but they can operate a lot of the equipment in the substation.

Batteries are charged during normal operation by battery chargers. Thus, they are always at peak and ready for emergency duty.

POWER DISTRIBUTION

Between the substation and the customer's lights, appliances, and other loads is the power distribution system. The lines that radiate from the distribution station are called feeder lines.

The feeder lines will be run either overhead or underground. Often two feeders are used side by side to serve the same load centers. Only one of the two is connected at any one time. The spare feeder is just that. In case of feeder trouble, feeder #1 can be shut down and feeder #2 used.

Both feeders have circuit breakers and switches. The two circuit breakers are kept closed. Only one switch is kept closed. The other switch is in the open position. In case of trouble in a feeder, the circuit breaker in that feeder can open. The other switch can be closed and service will continue without interruption.

The feeders connect to load centers. At the load center the feeders are divided into a number of primary mains. The load center has switches and fuses for each primary main.

The voltage at the load center is kept constant. As night approaches, or as days become hot or cold, lights, air conditioning and electric heat are switched on. These loads draw a lot of current and tend to pull down the incoming voltage. Yet the utility company maintains a steady voltage at the load center no matter what loads come on. The voltage is held steady by feeder voltage regulators at the substation. The voltage at the load is not allowed to vary, if at all possible.

The importance of the load centers owes to the load center's transformer. The primary mains connect the load center output to the distribution transformers. The distribution transformers are the familiar pots on the utility poles, or the transformers on concrete pads in backyards when underground wiring is used.

DISTRIBUTION TO CUSTOMERS

When the power company lays out a distribution system, the lengths of the mains are kept as short as possible. As the current flows along the mains, the voltage gradually drops. Between the substation and the load center transformer, the voltage loses a few percent. Between the load center transformer and the distribution transformer (along the primary main), the voltage again drops a bit. Between the distribution transformer and the customer's service entrance, the voltage drops a bit more (Fig. 3).

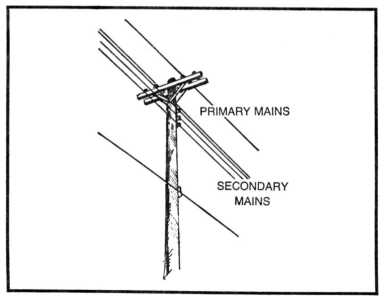

Fig. 3. The primary mains are usually run atop the distribution poles on the horizontal bars. The secondary mains are run below the primaries on the vertical pole.

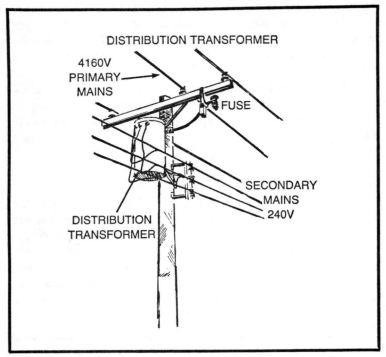

Fig. 4. On poles with distribution transformers contained in "pots," the transformers are wired between the mains and are fused.

The shorter the length of the mains, the smaller the voltage drop. Voltage drop is a dead loss to everyone concerned. The mains are constructed of large size wires to keep voltage drop to an economic minimum.

Between the primary mains and the secondary mains are the distribution transformers (Fig. 4). The distribution transformers receive their input from the primary mains. The voltage entering the transformers is usually 4160/2400 for residential or small commercial customers. The voltage leaving the transformer and traveling to the residential consumer is stepped down to the familiar 240/120 level.

For the consumers of large amounts of power, the transformer can receive 13,200/7,620 volts. The transformer can also step down this size of voltage for the consumer.

The lines leaving the distribution transformer and passing down the streets and alleys of residential neighborhoods are called secondary mains. From the secondary mains to the customer's service entrance, two-wire, three-wire and four-wire service lines are run, according to the need (Fig. 5).

VOLTAGES SUPPLIED

There are a number of voltage variations the utility company can supply to customers. Commonly they are two-wire, three-wire and four-wire services and various amounts of voltage.

The oldest type of residential voltage is 120-volt, 2-wire. It is rarely used anymore, except in small rural buildings and in other buildings which require only enough power for some lights and receptacles (Fig. 6).

The commonest supply is the three-wire, 120/240-volt, single-phase supply. It is actually a 240-volt supply which is connected to ground at its center point, splitting the 240 volts into two 120-volt supplies. Connecting to either end wire and ground provides 120 volts. Connecting across the two end wires provides 240 volts. These voltages are often 110/220 or 115/230. For most intents, the electrician considers all three of them identical, even though they are actually slightly different.

Commercial and industrial customers need three-phase service to power three-phase motors, ovens and other three-phase equip-

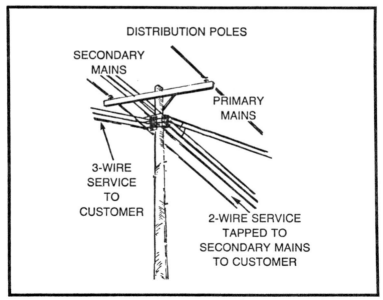

Fig. 5. Services are tapped from the secondary mains. Two- and three-wire service drops are run through the air to customers.

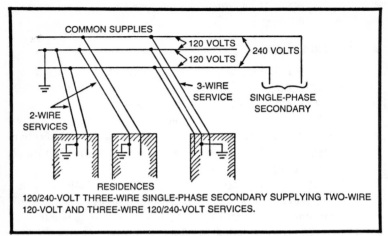

COMMON SUPPLIES

120 VOLTS
120 VOLTS
240 VOLTS

3-WIRE
SERVICE

SINGLE-PHASE
SECONDARY

2-WIRE
SERVICES

RESIDENCES

120/240-VOLT THREE-WIRE SINGLE-PHASE SECONDARY SUPPLYING TWO-WIRE
120-VOLT AND THREE-WIRE 120/240-VOLT SERVICES.

Fig. 6. Common supplies are 120/240-volt, 3-wire and 120-volt, 2-wire, single-phase services from secondary mains to residences.

ment. The three phases are sent out over three separate lines (Fig. 7). Each phase has its own line. Each phase is a separate source of electricity. A three-phase motor actually receives from three separate sources. Further information about this appears later in the book.

In addition to the need for three-phase power for large loads, these customers need single-phase power for lighting, receptacles and other small loads.

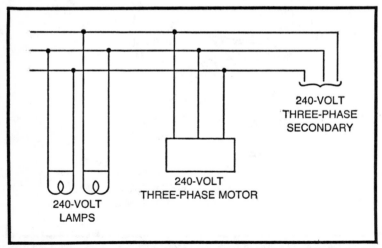

240-VOLT
THREE-PHASE
SECONDARY

240-VOLT
THREE-PHASE MOTOR

240-VOLT
LAMPS

Fig. 7. A commercial customer could receiver 3-wire, 3-phase, 240-volt service, wired as shown.

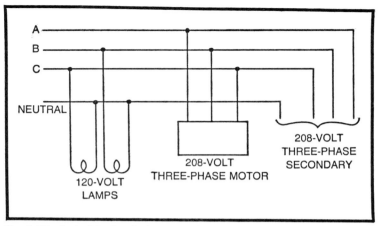

Fig. 8. The typical 4-wire, 3-phase supply provides 12-volt, single-phase and 208-volt, 3-phase service.

Each phase can be used as a single-phase supply. The utility can supply the three phases over a three-wire or four-wire system. The three-wire system provides 240 volts only. There is no 120-volt supply. If the customer needs 120 volts, he must have one of the phase lines centertapped to ground. That converts one of the phase lines to a three-wire, 120/240-volt system.

The four-wire supply provides 120 volts. The fourth wire is grounded and is called neutral. Between any of the three phase lines and neutral, 120 volts exists. However, the four-wire supply does not provide 240 volts. Between the phase lines there is only 208 volts. The supply is called 120/208-volt, three-phase (Fig. 8).

THE ELECTRICAL WIRING JOB

When an electrician is handed the responsibility of carrying out a complete wiring job, he brings to the project his own experience plus nearly a century of experience by his predecessors. Installing a wiring system is nothing new. This chapter speaks directly to the careerist electrician and to all who would become one, but the contents are relevant to the interests of everyone who engages in the installation of electrical wiring.

ELECTRICIANS AND ELECTRONIC TECHNICIANS

Be aware that when the electrician walks onto the job site, no electrical system exists. He must build one from scratch. Every wire, every piece of conduit, every fitting and device must be put into place. It is a demanding job.

This situation is unlike that of the electronic technician who works on equipment, already in existence, which was produced on an assembly line. This is why the average electrician is compensated at a higher rate than the average electronic technician.

The electronic technician's main area of attention is the workings of a circuit. He concerns himself with the way a signal is processed and passed through complex circuiting. Most of the time the tech is looking for trouble that consists of a single component or connection failure.

The electrician's central interest is the correct way to assemble conductors, conduits, fittings and devices into complete electrical systems. This is an entirely different job. The construction of the electrical system is a careful matter. It involves life and death. There is little room for error. An electrical system *must* be installed correctly.

FOUR KINDS OF WIRING JOBS

1. Troubleshooting, maintaining and updating wiring systems are common types of electrical work. Many electricians specialize in this type of work and rarely ao any other kind of electrical work. This is a house-call sphere of activity and there is always plenty of business. Most jobs are routine and easy, but occasionally some very difficult ones are encountered and considerable skill is required.

2. Wiring new homes is another common type of electrical work. Many such contracts are available when the homebuilding business is good. Small homes require small systems but larger homes often necessitate extensive and complex wiring that can rival many commercial and industrial jobs. Fitting out a multi-acred estate can entail an electrical system of large proportions.

3. Commercial wiring jobs consist of anything from wiring a small store to a small office building. A small store could receive a wiring job that is on a smaller scale than that of an average house. However, there are considerations in the store that are not part of a home wiring job, and vice versa.

4. An industrial wiring job can be routine or massive and complex, according to the application. Industrial jobs are not usually handled by small electrical contractors or independent electricians. Electrical engineering staffs and experienced, well-paid electricians work on these projects.

All of the wiring jobs must be safe. They all use similar methods and materials.

THE TYPICAL WIRING PROCEDURE

Whatever the type of electrical wiring job, the same basic techniques, materials and equipment are used. For example, an electrician walks onto a job armed with requirements, and from these requirements must first devise then install the system. To be more specific, the job could be a residential one and the electrician knows that the electric utility company is going to supply him with 120/240-volt, 60-cycle, single-phase current.

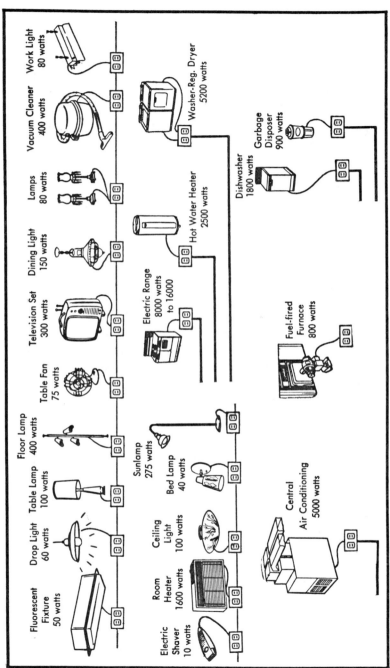

Fig. 1-1. The requirements of an electrical job are based around the loads that are to be energized.

The requirements he is armed with are the wattage loads that the electricity must handle. The loads can be the lighting, receptacles, garbage disposer, dishwasher, electric range, clothes dryer, air conditioner, and heater (Fig. 1-1). The electrician's task is to design and then install the system that will deliver the 120/240 volts from the utility company to the loads. The wiring system consists of a number of components (Fig. 1-2). They are the service entrance equipment, the branch circuit equipment, the overcurrent protection, the grounding equipment and any lightning arrestors that might be needed.

These components must be installed, by law, according to the National Electrical Code and in conformance with local electrical codes, rules and regulations. All of the design and installation procedures are spelled out carefully and in detail in the NE Code book and in local codes and ordinances. The design consists of calculating and choosing the correct size conductors, conduits, over-current protection, and other important hardware.

The installation consists of purchasing approved materials and installing the hardware according to prescribed methods. In most metropolitan areas there is a local inspection department that is charged with the responsibility of making sure the electrician does the job according to code. If he does not, he will not obtain official approval. He must keep amending the job until it is finally satisfactory to the authorities.

In some localities there is no electrical inspection as such and the electrician is mostly on his own. There is probably some inspection by the local electrical utility company and also by the local Underwriters' Laboratories representatives. Whatever the degree of inspection, the electrician is responsible for adhering to authoritative specifications.

If a faulty installation shows negligence, the electrician could become the object of a lawsuit.

THE APPRENTICE ELECTRICIAN

The process by which the Apprentice learns his trade and advances to Master status has been handed down from earlier years. In different localities, the training takes different forms (Fig. 1-3).

In areas where unions are strong, it takes a lot of effort to get into a professional electrician training program. The unions keep the numbers of newcomers low and stretch out the duration of the

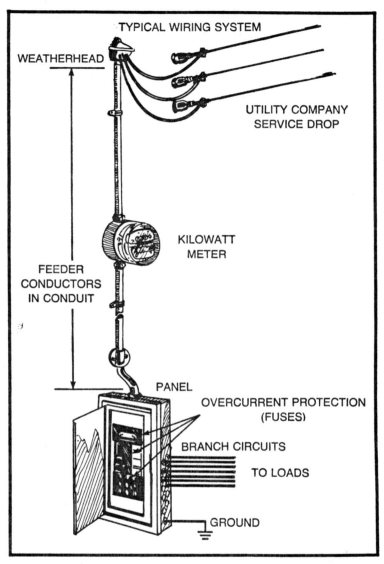

TYPICAL WIRING SYSTEM

WEATHERHEAD

UTILITY COMPANY
SERVICE DROP

KILOWATT
METER

FEEDER
CONDUCTORS
IN CONDUIT

PANEL

OVERCURRENT PROTECTION
(FUSES)

BRANCH CIRCUITS

TO LOADS

GROUND

Fig. 1-2. The components of a routine wiring system are the service entrance, branch circuits, overcurrent protection, etc.

Apprentice program. It can take as many as six years to pass through the Apprentice stage.

In localities where unions are not too active, the Apprentice can scoot quickly through the time allotted, according to his adaptability.

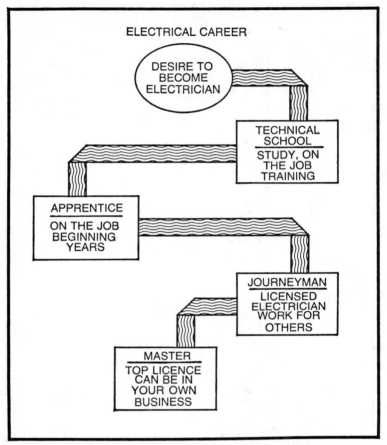

Fig. 1-3. From Apprentice to Master, the time-honored procedure for learning a trade.

The Apprentice is typically a helper. He mostly "pulls wire." This means that a Journeyman, Master, or other electrical expert appoints tasks to the Apprentice but supervises his work to avoid costly mistakes.

An Apprentice will be shown where a run of conduit is to be installed, will be given the material, and will hang it by the method he is told.

After the conduit is installed, the Apprentice will be given wires and instructed to pull them through the conduit. With the aid of fishing wire and other equipment he pulls the wires through the pipe he has installed.

There are many jobs an Apprentice does under supervision. He fetches equipment, pours concrete, installs wire without conduit, etc. What he does not do is calculate the size of the wires or calculate the size or type of pipe. His job is to help, and, as he does, he learns.

THE JOURNEYMAN ELECTRICIAN

As the Apprentice puts in his time, he should be engaged in a simultaneous study program. The program can be in a technical school, union classes, adult education classes, a correspondence school or one of self-learning. The study program is very effective when it parallels the apprentice's on-the-job training.

The study program is formulated in accordance with the tenets of the National Electrical Code. The study program should be geared to the goal of eventually taking the examination for a Journeyman's license. This license is a passport to many jobs and opportunities.

Most localities will not even let an applicant take the exam unless he can show a number of years of experience on the job. The experience can be accumulated in almost any way so long as it is legitimate, electrical vocation experience. The local jurisdictions want to be sure that the applicant can actually perform the physical operations of the trade in addition to passing the written examination.

The typical Journeyman's exam tests the applicant's knowledge of the NE Code book and his ability to understand and handle single-phase current. The exam might consist of two sections, with the first section consisting of 125 questions of the multiple choice variety. All the answers are in the NE Code book. You use the Code book to answer the questions.

This isn't as easy as it sounds. The applicant must be familiar with the Code book and be able to locate and interpret the pertinent rules and regulations.

The second section of a typical Journeyman's exam might consist of word problems. The applicant would be tested on the case problems he is likely to encounter in the field. He could expect to be given about six problems to solve. Here again, rules for the calculations can be found in the Code book. All of the calculations deal with the subject of sizing conductors, conduits, devices, and fittings.

Once he learns to ply his trade as an Apprentice, he should study at the same time he is gaining practical experience, then take his Journeyman's exam and pass it; then he is really on his way to a lucrative career.

THE MASTER ELECTRICAN

As the Journeyman works away, he has his sights set on becoming a Master Electrician. The Master Electrician is roundly respected and earns a high hourly wage, or he might even enter the ranks of the electrical contractors. The Master knows a great deal about electrical wiring and is given heavy responsibility. Of course, he is well paid to handle that responsibility.

The Master Electricians are an elite group. To earn a Master's ticket these men had to acquire a lot of experience and had to pass an exam that covered 3-phase electricity, transformer sizing, motor and feeder currents, and a large number of other formulas and calculations.

The Master Electrician becomes intimate with the Code book through study and experience. The Code book is written so that it encompasses nearly every possible situation that could occur. This is an unattainable goal since there are always electrical problems cropping up that are not covered in the Code book or else the book's recommendations aren't practical for a particular installation.

The Master Electrician, with the approval of the local jurisdiction, then works out a professional answer to the problem. He must be able to produce a solution which is satisfactory, safe, and exhibits top-flight workmanship.

The Master Electrician works at a level where many of his decisions are not questioned. He is his own man and makes most of the final decisions that go into a job. During an electrical wiring job, he is at the place where the "buck stops."

THE CODE CHAPTERS

To progress without undue delay from Apprentice to Master, you readers who wish to become professionals must learn the Code book. It is not necessary for you to memorize particular sections of the book; it is only necessary for you to know your way around the book.

On the job, under supervision the Apprentice will learn how to handle the electrician's tools, how to hang conduits and run conductors, how to install devices and fittings, and he'll become familiar with all the methods and materials.

The instruction which an Apprentice receives is based on the readings of the Code book. The chapters are laid out in the following way. Chapter 1 is a general discussion and contains definitions and requirements. The sections of the chapter are numbered sequentially, from 100 upwards (but no higher than 199). The other chapters are treated in a similar way; for instance, Chapter 3 is numbered from 300 to 384.

In Chapter 1 the words that are defined look familiar but they are very specific and have different meanings than words that look quite the same: for example, grounded conductor and grounding conductor. These are two different items in an electrical system. The grounded conductor is one of the wires in the system that carries current—such as the neutral wire. It is grounded.

The grounding conductor, on the other hand, is a length of wire that ordinarily does not carry current and which connects the frames of appliances to ground for safety purposes. The point is that the definitions must be learned exactly. As you read the Code, confusion is lowered to a minimum, if you understand the definitions.

The rest of Chapter 1 under Article 110 spells out general requirements of workmanship and materials.

Chapter 2, Wiring Design and Protection, treats the subjects thoroughly enough to cover most jobs. The chapter delves into the calculations needed to produce a safe wiring design. For Journeyman and Master Electrician exams, Chapter 2 gets a lot of attention.

Chapter 3 deals with Wiring Methods and Materials. Article 310 contains a number of Tables, starting with Table 310-16. After the Tables there are Notes. Tables and Notes are used constantly during most calculations. The Tables, in general, show the amount of current that can safely pass through various sizes and types of wire.

Chapter 4 covers Equipment for General Use. Cords, cables, wires, fixtures, appliances, heaters, motors, air conditioners, and batteries are discussed, along with other equipment.

Chapters 1, 2, 3, and 4 are the important parts of the Code which working electricians and inspectors use constantly.

Chapters 5, 6, 7, and 8 cover special occupancies, equipment, conditions, and systems. In the course of his daily tasks the career electrician might become involved with these areas of electrical work and become familiar with these chapters. While studying the Code for exams he must learn his way around these parts of the book. Hazardous locations, bulk storage areas, motion picture studios, and so forth are not always out of the realm of daily work. However, most of the electrical work engaged in during an average electrician's career is covered in Chapters 1, 2, 3, and 4.

Chapter 9 is full of Tables and has some examples of how to calculate typical jobs. The aspiring careerist will be using Chapter 9 constantly and it is important that he understand each of the examples thoroughly so that he can calculate according to Code.

CONDUCTOR AND CONDUIT CONSIDERATIONS

The Code calculations are a vital part of the knowledge an electrician must have. In many electrician's exams, the calculations deal mostly with sizing conductors, conduits, fittings, and devices.

Conductors carry current from the utility company *drop* to the panel. These conductors, according to the Code, are *feeders*. Fundamental to any electrical installation is the calculation of the size of the feeder conductors.

Conductors carry current from the panel to the various loads. The lighting, receptacles, appliances, and motors in an electrical installation are the loads. The conductors from the panel to a load are in a branch circuit. The size of the branch conductors is an important calculation that must be made according to Code.

Conductors such as Romex are so well insulated that they can be pulled from the utility company's drop to the loads. No conduit is needed. However, the use of conduit is a common method of wiring and some localities will not permit wiring without conduit.

Conduit comes in a number of sizes. According to the number of wires that must be run, a particular size of conduit has to be chosen. The Code specifies that you can only fill conduit up to a certain percentage. Conduit fill is an important part of daily electrical calculations.

OHM'S LAW

Ohm's Law is not mentioned in the Code book. Yet every article in the book has something or other to do with Ohm's Law. It is *the* electrical law. To an electrician it is just as important as the Law of Gravity.

The electronic technician is also concerned with Ohm's Law, but from a slightly different viewpoint. The electronic tech normally works with resistances in ohms and does not worry too much about power in watts.

The electrician, while needing to use resistance in ohms quite a bit, works more at calculating watts. The electrician uses a variation of Ohm's Law called the *Power Formula*.

There are more resistance-in-ohms calculations by electronic techs and more power-in-watts calculations by electricians. The utility company figures its power charges in watts. The unit used is the kilowatt hour, which is nothing more than power multiplied by time.

Ohm's Law, from the electrician's point of view, is discussed in Chapter 4 of this book. Ohm's Law must be second nature to the electrician. As he works on electrical circuits, the principles of Ohm's Law and the Power Formula should be ever on the electrician's mind.

LOADS

The wires the electrician pulls during an installation lead to the loads. The loads, as mentioned, consist of units like lights, receptacles, appliances, and motors. The amount of the loads determines the gage and length of the conductors to be installed.

Even though the load draws current through the conductors and consumes energy produced by the current, loads are not always measured directly in terms of the amount of current.

The electrician views load consumption in numbers of watts. The watt is the unit of measurement most convenient for the electrician to use.

When you calculate total loads, you add up watts. For example, to figure out the total load of a house, you add up the watts that are used in all the loads. Thousands of watts are used in an average house load so that it is more convenient to use the kilowatt (1000 watts) as a unit. This simplifies things by eliminating three digits at the end of the numbers.

The load of each unit is prominently displayed on it. For instance, a light bulb would have 150 watts stamped on the glass. A clothes dryer would have 4.5 kW stamped on a metal plate attached to the rear panel. A barbecue would have 1.8 kW stamped on a nameplate on its bottom. A motor, however, might have ½ horsepower stamped on its nameplate.

When you find a nameplate with a horsepower designation, you must convert the hp to kW if you want to add it directly to the kilowattage of the other loads.

A dishwasher could have 10 amperes stamped on its nameplate. To add the dishwaser load to the rest of the loads, you must first convert the amperes to kilowatts.

Calculating loads is one of the difficult and vital calculations an electrician must be able to perform. Every Journeyman's exam will likely contain a load problem, usually involving a residence with a single phase. A Master's exam might ask the applicant to calculate loads in a commercial establishment (like a restaurant) with 3-phase loads. Single-phase load calculations are discussed in Chapters 13 and 14 of this book. Commercial and industrial loads are treated in Chapters 16 and 17.

VOLTAGE DROP

After the load of an electrical installation is calculated, wire can be sized to carry the current to energize the load. In some instances the wire has to be run a long distance, perhaps a hundred feet or more. When the conductor has a long length like this, some of the current's energy is necessarily consumed and lost. The Code will not tolerate a voltage loss of more than 3 percent between the panel and the load.

When more than 3 percent is lost, the Code considers the voltage drop to be excessive. One way to reduce voltage drop is to increase the size of the wire. Therefore, the size of the wire that was calculated beforehand according to load must be further calculated to accommodate excessive voltage drop. This is discussed in greater detail in later pages.

AMBIENT TEMPERATURE

Besides the loss of energy when wire runs are too long, another factor has to be considered in calculations. When the temperature of locations where wire is to be run is 86°F or below, this factor can be ignored. But when the temperature rises *above* 86°F (or its counterpart, 30°C), the ambient temperature causes overheating in the wire and energy is lost.

The Code provides instructions and formulas for combatting the problem. The usual course of action is to increase wire size, as you must do when the voltage drop is excessive. These calculations—normally found in exams—are covered in the text.

BALANCING

After all the conduits and conductors are sized, there is one more major calculation that must be made on a wiring job. It's called balancing.

Just as a seesaw will not ride smoothly if one side has a heavy load and the other side a light load, an electrical installation does not operate efficiently if one side has a heavier load than the other.

There are two sides in a 120/240-volt, 3-wire resisdence job. Each must carry the same, or nearly the same, load. If not, conductors can overheat, fuses can blow, and other inefficiencies occur.

No electrical wiring installation is ever perfectly balanced, but a careful distribution of the kilowatts is necessary. The Code dictates that all circuits be balanced at installation. It is standard practice to consider balancing in every wiring calculation.

THE COMPLETE ELECTRICIAN

Becoming an electrician and performing wiring jobs correctly takes mechanical dexterity and knowledge. The electrician uses his hands and his head. The hands have skill with tools and the head directs the activity—with the National Electrical Code as the main guide.

The electrician must be expert enough to understand each of the conductors, conduits, devices, and fittings. He connects the hardware into circuit components that in combination form a complete system. He grounds the equipment and ties to ground the set of conductors. He comprehends the way the electrical circuits work and the function of each component and he knows what terminals the wire ends attach to. He installs these systems in residences, barns, stores, and factories.

This book discusses—from the author's own point of view—tools, basic theory, conduits, conductors, fittings, devices, components, grounding, circuit connections, and the various kinds of installations an electrician works with.

If you are an electronic technician, this text should widen the scope of your abilities. If you are interested in becoming a professional electrician, this book will teach you a great deal about the job you're trying to learn. If you are a do-it-yourselfer, you'll find that the serious level of instruction within these pages will advance you well along your way to becoming a competent and versatile electrical wireman.

ELECTRICITY SAFETY FIRST

In any wiring project safety is every bit as important as the work itself. To the worker safety should be a way of life.

TRAGEDY

There are many cases every year like the following.

On a hot summer night in Chicago, a homeowner was astonished to see his whirring electric fan suddenly burst into flames (Fig. 2-1). The fan was sitting on the window sill and the drapes caught fire immediately. In no time at all the window was a sheet of flame.

Undaunted, the homeowner ran into the yard, turned on the garden hose, and directed a stream of water at the burning window. As the water struck the flames, the man collapsed. He was dead.

The stream of water had fallen onto the burning fan, which was still plugged into the electric receptacle. The water served as a conductor and conveyed a lethal current of electricity into the unlucky man's body.

Standing on wet earth, he was electrocuted.

Another common type of electrical tragedy unfolds in much the same manner as the following incident. In a remote section of Georgia, there is little or no formal regulation of electrical installation procedures. As a result property owners are free to operate with whatever regard for safety they choose.

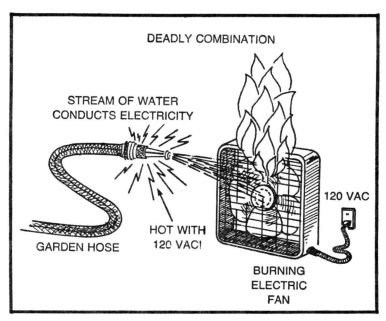

DEADLY COMBINATION

STREAM OF WATER
CONDUCTS ELECTRICITY

HOT WITH
120 VAC!

GARDEN HOSE

120 VAC

BURNING
ELECTRIC
FAN

Fig. 2-1. Do not try to put out an electrical fire with a hose, the lethal energy is liable to travel to you along the stream of water.

A landowner was doing some re-wiring in his barn. He had purchased a new milking machine and was running a heavy duty line to service it. He had a fairly good knowledge of electricity, sufficient at least for his own purposes.

At the switchbox he threw the main switch to OFF, and began running cable. The run was a long one and he snaked the cable step by step from the panel to the site of the new milker. For illumination he used a battery-powered, portable lamp.

As the job neared completion, the man crawled under the milker and began making the 240-volt connections. Blammo! Without warning the electricity came on and the man was found electrocuted—a hot line clamped in each fist.

A hired hand had entered the barn, found it dark and threw on the main switch.

Similar cases occur regularly. Mostly they happen to people who are inexperienced, or who do not follow safety rules. Rarely is an experienced electrician involved.

The homeowner with the burning fan should not have died. Death would have been prevented if he had pulled the plug on the fan

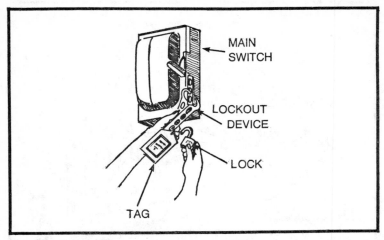

MAIN SWITCH

LOCKOUT DEVICE

LOCK

TAG

Fig. 2-2. An experienced electrician always uses a lockout device to prevent accidental activation of circuits that are being worked on.

before attempting to put the fire out with water. If he couldn't get to the plug, then he should have turned off the main switch of the home before putting water on an electrical fire.

The Georgia farmer would still be alive if he had followed a simple safety procedure that is automatic to electricians. If the main panel is switched to OFF and an electrician doesn't want it tampered with, he installs a lockout device (Fig. 2-2). This is a padlock affixed with a tag which reads DO NOT TOUCH. With the padlock on, even if someone ignores the warning sign, the switch can't be thrown.

THE NATIONAL ELECTRICAL CODE

The National Electrical Code is the basis on which all of our nation's electrical work is performed. Every professional electrician should know intimately the NE Code. Non-professionals would benefit similarly by becoming familiar with the Code. It is contained in a carefully written book that is prepared every three or four years by hundreds upon hundreds of authors (Fig. 2-3). The book is as exact as is humanly possible.

Those who consult it for the first time find the book difficult to understand and make little headway in applying its wisdom on an actual job. The Code must be carefully studied, even though the nuances are subtle and interpretation is laborious.

Even the experienced electrician carries a copy of the latest Code with him on jobs. The copy is dog eared and cluttered with penciled notes. The inspectors also carry Code books with the same well-worn look.

One of the main purposes of the Code is SAFETY. This is plainly stated in Article 90-1. Safety pertains foremost to the protection of people. Secondly, safety pertains to the protection of property.

Article 90-2 of the Code defines the type of installations it covers. This book, like the NE Code book, is concerned with public and private buildings such as homes, stores and such structures.

The Code, because of its preoccupation with safety, is not at its best as a design manual, nor is it always efficient, convenient and adequate. It simply tells you what the writers of the manual feel is the safest possible procedure. A common misconception is that if the

Fig. 2-3. The Code book is the electrician's rule book and has as its primary concern safety for people and property.

job is done according to the Code, it is satisfactory in all respects. It's not! It is safe, that's all.

Without dispute, safety *is* important, but an electrical installation that takes into consideration new technical developments and provides for future expansion of the system is also important. Furthermore, it is a good idea to build into an installation a margin of safety that will offer additional protective features.

The Code gives you minimum provisions. The Code will give you an installation that is "essentially free from hazard," but not the best of all possible installations for your situation.

This brings the dollar into the picture. A large percentage of electrical jobs are performed strictly to Code. If you can interpret the Code book well, and if you follow its prescriptions carefully, you'll be able to produce a safe installation which is also inexpensive.

There won't be a piece of conduit an inch longer than necessary; there won't be a conductor a size longer than necessary; and the service panel will not have an empty space; and so on. If you bid on a job on this basis, you stand a good chance of underbidding everyone.

However, if a single large appliance is added in the future, the system might not be able to handle the load. If this should happen, the person who paid for the work would be very dissatisfied with the adequacy of the system.

To avoid such a situation, most electricians follow the Code for safety but produce an installation design of their own which will meet future expansion needs and which can easily be readapted for system changes.

This, of course, costs the client additional money, but you should persuade him to give you the option of installing whatever you know to be necessary to accommodate immediate loads and meet possible future needs as well. This will entail building into the system even more safety measures than the Code advises.

CODE ENFORCEMENT

In Article 90—the introduction—of the Code, there is a section (90-4) on Enforcement. This part informs the reader that the Code book is not enforced by the authors of the Code themselves.

It starts off by stating: "This Code is intended to be suitable for mandatory application by governmental bodies...." Notice that it does not say this Code *shall* be obeyed.

Section 90-4 turns the enforcement of the Code over to "...the authority having jurisdiction of enforcement of the Code." This means the local inspectors, utility companies, townships, cities, counties, etc. The local authorities have the last word. Even if a job does not comply with the Code's recommendations, a local inspector could still O.K. it.

The Code is a superguide. It spells out in no uncertain terms what the authors believe to be the safest way to do the job. Then it is up to the local jurisdiction to make sure the job is right.

Many local Codes and ordinances run contrary to the NE Code. The local Codes take precedence. If the locals authorize it, then it *can* be installed. It could possibly be wrong, but, after all, it is the local jurisdiction you must work with.

Fortunately, the NE Code is in most cases the ultimate authority in any dispute over wiring technique. But remember that Section 90-4 says that "the inspector is always right."

THE NFPA

The National Fire Protection Association is a wonderful nonprofit organization with one objective in mind: the protection of lives and property from fire.

The Association does this by aiding in research "for better methods to harness the energies of fire." We need fire. Our civilization would not be the same without fire—that is, fire under control. Out of control, as we all know, fire is one of the Four Horsemen of the Apocalypse.

Many fire hazards are around us. There is no need to go into them. The NFPA investigates every source of hazard. The NFPA has a section that concentrates on electrical safety.

The NFPA sponsors the Code (Fig. 2-4). The Code was originated in 1897 as electricity became popular. Insurance, electrical, architectural and other allied interests produced the first Code.

In 1911 the NFPA took over the sponsorship of the Code. The organization has been producing and updating the Code books ever since. In fact, the Code is really NFPA's National Electrical Code.

Fig 2-4. The Code book is prepared by the Electrical Section of the National Fire Prevention Association.

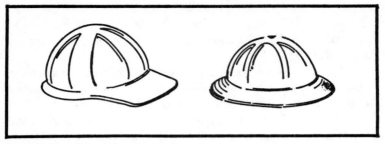

Fig. 2-5. Two styles of hard hats.

The NFPA is the Daddy of a lot of the fire prevention Codes and standards. The federal government has enacted legislation according to Codes and standards produced by the NFPA. One Act of interest to electricians is OSHA.

OSHA

The Occupational Safety and Health Administration was formed by an Act of Congress to set standards to keep workers healthy.

OSHA reaches into all industry and on all levels, from the floor of an open hearth furnace to the executive suite. OSHA has adopted the NE Code. This brings the Code under the law of the land.

OSHA has one important capability that the NFPA lacks. It has policing powers and whoever breaks or bends the law is subject to enforcement of the law by OSHA.

A lot of the Codes and standards of the NFPA have been adopted by OSHA. OSHA approves various safety items. When you see a designation of OSHA approval on a pair of gloves, shoes or other safety clothing, you know that the product is safe to use.

As far as electrical work is concerned, OSHA overhangs the safety field and you can count on it as a strong ally. To learn more about OSHA, you can write to the National Institute for Occupational Safety and Health.

ELECTRICIAN'S DRESS STANDARDS

Every kind of worker dresses to fit the job. A professional electrician does too. When he enters a home, the customer sees him not as a man but as an electrician.

Naturally he wants to look presentable; but the main reason for the electrican's style of clothing is safety. The prime consideration is to insulate himself in case he contacts electricity.

Starting from the top down, he can wear a hard hat (Fig. 2-5). It's made of an insulating material and is called a Class B hard hat. It will protect him if a hammer falls on his head and it will not conduct electricity from any hot line that he might accidentally touch.

Below the hat he could wear a pair of safety goggles. Should particles start to fly, his eyes will be protected (Fig. 2-6). Also, he'll be wearing a long sleeve shirt, even if it is hot.

When working on a live circuit, he'll put on a pair of rubber gloves. In addition he'll don a pair of oversize leather gloves over the rubber gloves. The leather gloves will prevent the rubber gloves from being torn by some sharp point on a wire (this is especially important if the wire is a hot wire).

The electrician wears no jewelry. Rings, wrist watches, bracelets, metal belt buckles, exposed zippers, metal buttons, and so on are taboo. Any of them could become an electrode against his skin in the wrong circumstances.

His shoes will always have rubber soles and heels—the thicker the better, as far as electricity goes. They shouldn't be so high though that they cause a loss of balance (Fig. 2-7a).

His shoe should have the old steel toe. The steel toe can withstand a conduit smacking down on top of it. His own toes will be saved (Fig. 2-7b).

In certain job environments, all kinds of protective clothing will be needed. There are OSHA approved hair coverings, ear protectors, face shields, welder's helmets, hand pads, knee pads, special

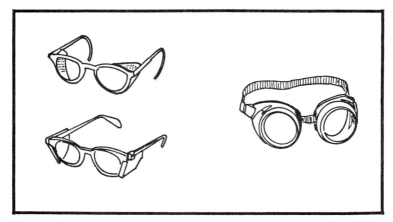

Fig. 2-6. Safety glasses and goggles protect the electrician's eyes from flying particles.

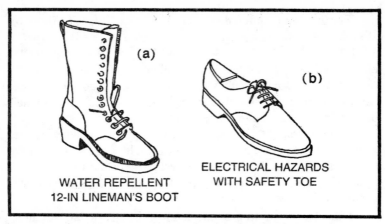

(a)

(b)

WATER REPELLENT
12-IN LINEMAN'S BOOT

ELECTRICAL HAZARDS
WITH SAFETY TOE

Fig. 2-7. (a) Safety boot with insulated sole and heel and steel toe. (b) This oxford has the same safety features as the boot.

insulated sleeves, protective creams, foot guards, oxygen-supplied helmets and full suits, etc.

Clothes do make the man (or woman) and this is even true of electricians. When he goes to work, he suits up to fit the job. His clothes should have a lot of deep pockets and not tear easily. They should be snug so they don't snag on anything, yet loose enough to allow lifting and squirming easily.

The electrical worker should always be dressed warmly enough, yet not overly so because he'll be sweating enough as it is. The idea is to make himself as safe and as maneuverable as possible. Be comfortable and stay healthy.

MATERIALS STANDARDS

The Code tells you how to install materials in the right way. What about materials? How do you know they *are* the right materials?

The right materials are "listed by Underwriters." You've heard, of course, of the Underwriters' Laboratories, Inc. (Fig. 2-8). The fact that it is a not-for-profit organization doesn't mean it doesn't make money. It does, since it must have income to provide its services. However, its basic purpose is not to make money; its purpose is to establish safety.

When a manufacturer produces an electrical product, he submits a sample of the material to UL. He pays for the laboratories to test the material. If the material meets safety standards, it earns a UL listing.

Listing by the UL doesn't mean that a material is better than some competitive material. It means the material has passed

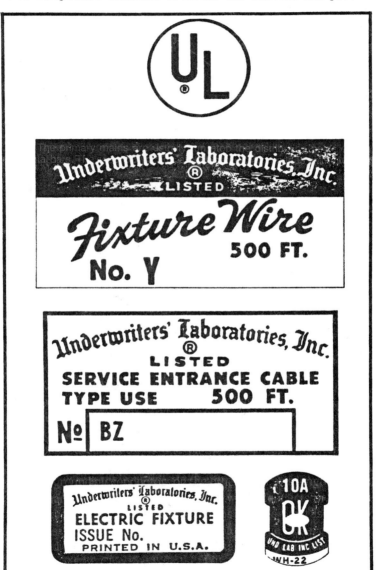

Fig. 2-8. Typical labels of approval by the Underwriters' Laboratories, Inc.

minimum safety standards. It will be safe on the job it was designed for. It won't be safe if it is overloaded on a job it was not designed to handle.

With materials you still only get what you pay for. A UL listing does not imply good quality; it merely indicates that minimum safety standards have been met, whether the item will last a long time or not.

The listing by UL is subject to a "follow-up service." This means that UL can regularly inspect manufacturing plants and test products purchased on the open market, and otherwise keep an eye on the material. If the safety standards of the material should slip, the UL listing would be rescinded and the manufacturer would not be permitted to mark UL Listed on his product.

On approved material, the manufacturer is allowed to install a Listing Mark. On small products like sockets, receptacles, outlet boxes, and snap switches, the familiar symbol of the circle enclosing the letters UL is stamped on the product along with the manufacturer's name. On larger products like wire, cable, panels, lighting fixtures, and conduit, an Underwriters' label is pasted or tagged (Fig. 2-9). The UL symbol in the circle might also appear on the nameplate, along with other pertinent information.

A large number of materials on the market are not listed by Underwriters'. Even though these materials might be safe, it is easiest to simply purchase only listed materials. The only exception to this general rule would occur in the purchase of low voltage items like doorbells. The UL doesn't bother with them since the low voltage presents no hazard. The doorbell transformer though *is* listed since its input is at the regular line voltage.

Most electric motors are not listed. You'll have to be careful when purchasing them.

Fig. 2-9.To earn this tag a manufacturer must submit his product and let it be tested for safety.

ELECTRIC SHOCK DANGER

During the installation of electrical systems a professional will be working with live wires a very small percentage of the time. Most often he'll complete the job and leave and a couple of days later the utility company men will test the installation. If no defect is found, they'll turn on the current. The professional might never see his job operating.

During maintenance and troubleshooting, you as the professional *will* work with live systems. When you actually do the wiring, however, you'll disconnect the system from the line current. You won't reconnect until your job is completed, unless some instrument testing is required.

Yet there will be some occasions to work on live systems. During these times it is possible that you will be exposed to electric shock. In addition, emergency situations sometimes arise. As an electrician you should know as much as possible about electric shock. You might find yourself in a situation where someone's life would depend on your knowledge of electric shock. That someone might even be yourself.

It is a fact that the old timers do not get shocked often. Only the novice might ignorantly place himself in danger.

The best safety precaution is to know your way around. You'll gradually develop good natural safety habits.

BODY STANCE

As you get accustomed to your work and become familiar with the materials, you'll develop the body stance of the professional electrician.

Your tool pouch will be suspended below your waist on one side and your gloves will be tucked into a pocket. Your tool box will sit at a certain location and your movements will be sure and accompanied by no wasted motion.

All this adds up to safety. You are in a protective body position ready to react to conditions.

The main enemy you protect yourself from is electric shock. There will be occasional jolts but your body stance practically eliminates your getting hurt.

For you know that electrons travel in a closed circuit and you try not to be part of any accidental circuit. If you *do* become part of a circuit, you do not let the current flow pass near your heart.

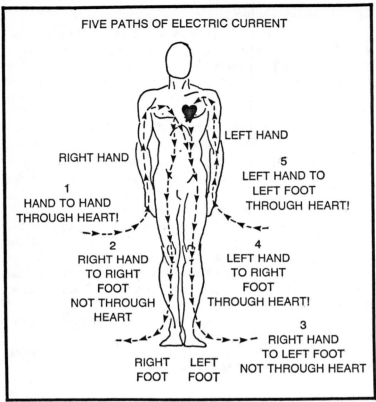

Fig. 2-10. Electric current can take any one of these five paths when the hands and feet are involved.

The current can commonly find a number of paths through your body. The paths near your heart are dangerous. Paths through your body far from the heart can be frightening but are not serious.

You touch, for the most part, with your hands and feet. This makes the following paths possible: hand to hand, right hand to right foot, right hand to left foot, left hand to right foot, and left hand to left foot (Fig. 2-10).

If 120 volts AC should pass along any of these paths for even as little time as two seconds, internal tissue can be heated or burned. Should one of the paths be through the heart, the electrical impulse that controls the heart beat can be disrupted. The heart will cease its rhythmic beat and begin to flutter. The heart will not be able to pump the blood properly. This condition is called ventricular fibrillation (Fig. 2-11).

The burning of tissue is usually not serious and healing occurs in the same way as it does with any other burn. The fibrillation *is* serious. If it is not stopped quickly and the regular heart beat restored, death occurs. Most emergency vehicles are fitted with de-fibrillators, also called paddles. These devices apply an electrical shock to the body to restart the regular heart beat.

The best safety policy though is to prevent the current of an accidental shock from ever getting near the heart. All electrical workers know to keep their left hand in a pocket whenever possible

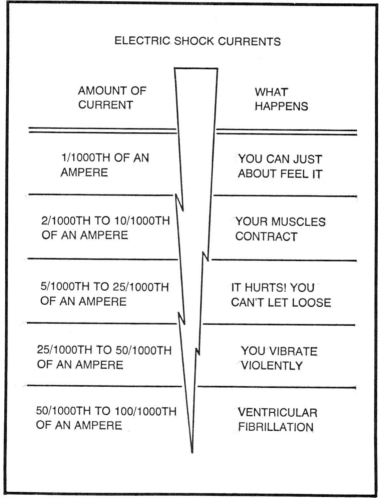

ELECTRIC SHOCK CURRENTS	
AMOUNT OF CURRENT	WHAT HAPPENS
1/1000TH OF AN AMPERE	YOU CAN JUST ABOUT FEEL IT
2/1000TH TO 10/1000TH OF AN AMPERE	YOUR MUSCLES CONTRACT
5/1000TH TO 25/1000TH OF AN AMPERE	IT HURTS! YOU CAN'T LET LOOSE
25/1000TH TO 50/1000TH OF AN AMPERE	YOU VIBRATE VIOLENTLY
50/1000TH TO 100/1000TH OF AN AMPERE	VENTRICULAR FIBRILLATION

Fig. 2-11. The degree of harm that can happen to you depends on how much current actually enters your body.

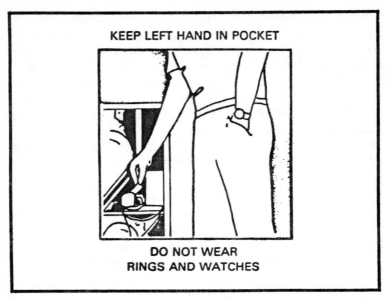

KEEP LEFT HAND IN POCKET

DO NOT WEAR
RINGS AND WATCHES

Fig. 2-12. An electrical worker knows that if his left hand is in his pocket, the path through the heart is eliminated.

(Fig. 2-12). Current can't enter or leave through the left arm if the arm is not a part of the circuit.

The current in the electrical system can flow from a hot line to ground or from one hot line to another at the other end of the system. As you learn the circuits you'll know which lines are which. That is your best protection.

Your experienced body stance will automatically keep you clear of the hot lines and the grounds.

FIRST AID FOR ELECTRICAL SHOCK

Anyone working with electrical wiring should inform himself about electrical shock. During an emergency you may be called upon. If someone in your neighborhood gets badly shocked, people will run to you for help.

A typical situation where you might be asked to render aid happens during severe lightning and wind storms. Either lightning or wind could knock down a utility pole and strew exposed live wires about.

One such storm knocked down a pole and a broken wire hit a passerby. The man fell to the ground. A second passerby came to his aid and tried to pull him from the wire. The second man fell too. The pair of victims lay askew—the electric crackling all the while.

At that moment the utility company's truck, searching for the cause of the power outage, arrived on the scene. Two electricians leaped out and peeled off their heavy lineman's coats. Avoiding the spitting wires, each worker went to a victim and wrapped the stricken man with his lineman's coat. It was obvious to all witnesses that the utility workers were carefully trained.

The workers were watching the hot wires with one eye and assisting the injured with the other. They did not touch the pockets or shoes of the victims, since a pocketful of metal change or nails in the shoes could easily transmit the charge to the rescuers.

A second service truck rolled up and crisp orders were broadcast over the two way radio. The hot wires lying on the ground suddenly stopped sparking. The linemen bent over the victims and applied mouth to mouth resuscitation.

The victims began breathing again. Warm blankets were wrapped around them. An ambulance arrived and sped them to the hospital.

The knowledge and quick action of the linemen were instrumental in saving two lives.

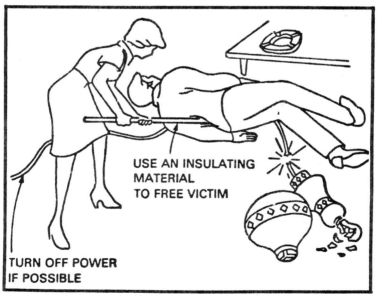

USE AN INSULATING
MATERIAL
TO FREE VICTIM

TURN OFF POWER
IF POSSIBLE

Fig. 2-13. In the home a broom stick can be used as an insulator to disengage a victim from a hot line.

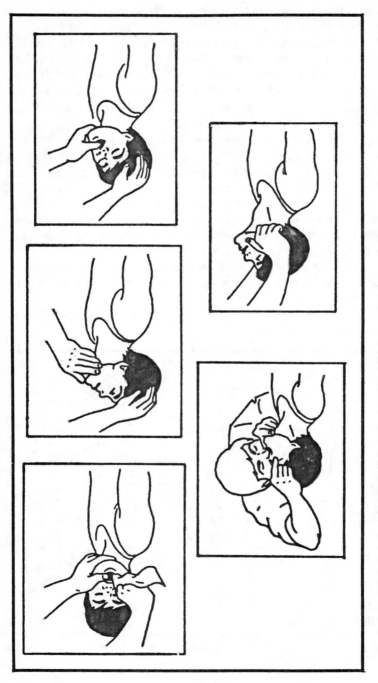

Fig. 2-14. Learn this step-by-step resuscitation measure. It can possibly save a person who has been electrocuted.

ELECTRIC SHOCK TECHNIQUES

The first aid procedure for electric shock is not complicated and can be executed quickly. First of all, if a victim is still in contact with a hot wire, you must disengage him.

An insulator has to be used. The linemen used their heavy coats. Another handy item usually available is a wooden stick; a broom, for example. Electricity won't travel along the wooden handle. Push the victim away from the electricity with this or some other insulator (Fig. 2-13).

If the victim isn't breathing when he is released, you must apply artificial respiration (Fig. 2-14). The best technique is mouth to mouth resuscitation. It is done with the following steps:

1. Turn the victim's head to the side and feel in the mouth for objects like false teeth, chewing tobacco, gum, etc. If there is any, remove it; otherwise it could end up down his throat.

2. Straighten the head and tilt it back so the chin points upwards.

3. Force your left thumb between the teeth. Raise the lower jaw above the upper one.

4. Hold the victim's nose tightly closed. Take a deep breath and place your mouth completely over the victim's mouth so that no air can escape between. Begin blowing forceful breaths into his mouth. (If it's a child, blow gently.) Blow 20 times a minute for adults and 12 times a minute for children.

5. If you can't get the victim's mouth open, extend his head and hold the lower jaw firmly with both hands.

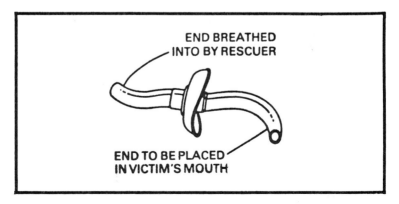

END BREATHED INTO BY RESCUER

END TO BE PLACED IN VICTIM'S MOUTH

Fig. 2-15. In your tool box or truck keep a mouth-to-mouth plastic aid to make resuscitation measures easier to perform.

There are a number of plastic devices to aid in administering artificial respiration. As an insurance measure, even though you'll probably never need it, get one of these devices and keep it available. It might someday save a life (Fig. 2-15).

Your concern for safety should never relent—not for a moment. You as an electrician should practice safety as part of your daily work program. Safety is all-important and should be given your lifelong interest.

Approach to Tools of the Trade

An electrician's tools are indispensable to the best conduct of his work. He respects these implements of his trade and he uses them with precision. Whoever would become an electrician must know the tools and techniques and learn to put them quickly and deftly into practice.

TOOL PLANNING

If you watch an expert electrician work, there is little noise or excitement. The electrical system almost magically takes shape or is repaired.

Entire systems are installed long before the power company arrives on the scene to turn on the current. Quite often the electricians are gone when the electricity actually begins flowing in the system and the installation lights up.

Most of the time, the new system works perfectly from the instant it starts up.

As you watch the competent electrician work, you'll observe that when he pulls a screwdriver from a pocket of his tool pouch (Fig. 3-1), he returns the screwdriver to the same pocket. That old expression keeps coming to mind: "a place for everything and everything in its place."

If you watch a not-so-expert electrician, you'll see that after a few minutes on the job tools and hardware are scattered all over the place. When a tool is needed a second time, a search halts the job

Fig. 3-1. An electrician carries his most used tools in a pouch that is specially made to wear on the hip.

until the tool is finally located at the bottom of a pile of equipment and materials.

Every extra move you make, every tool hunt you have to conduct, every bit of sloppy technique costs time and money. It has been shown that a neat, organized worker can do an electrical job in half the time required by a slovenly worker.

From the very first, establish correct work habits.

TYPES OF TOOLS

The electrician is a construction worker. He is expected to know and be skillful with screwdrivers, hammers, pliers, wrenches, saws, ladders and electric drills.

In addition, the electrician must be able to make professional electrical connections, to fish wire through conduit, and to assemble several kinds of pipe. The next few sections of this chapter discuss some of the details for making wire connections. Chapter 5 (Conduit) covers the installation of pipe and some necessary techniques. Chapter 6 (Conductors) provides more information on wiring techniques.

Another "tool" of the electrician is a certain amount of basic theory (from a practical point of view). The theory enables him to visualize what is happening inside the electrical system, how to intelligently use meters and instruments and how to read work drawings. These aspects of theory are discussed later in this chapter.

Theory itself in the form of Ohm's Law and the Power Formula follows in Chapter 4. Theory for the electrician must be related closely to the NE Code book.

The expert electrician is expert with the Code book. He performs his work according to Code. There isn't another tool in the electrician's truck that has as much importance as the Code book. The Code looms above this entire text. Every move on the job is foreshadowed by the Code.

Arrange your tool pouch and tool box so that tools and testers get the easiest of handling. Every important tool should be at your finger tips, at least the most frequently used tools.

A little tool planning right now, before you proceed any further, will pay handsome dividends in the future.

ELECTRICAL CONNECTIONS

Wires must be connected in electrical systems. That's obvious, yet often connections are poorly made. When that happens there is danger of eventual costly repairs, and of fire-explosion.

Good connections are made by experienced electricians. Bad connections are not made by experienced electricians. To make connections there is one right way and many wrong ways.

For example, let's consider removing insulation. The best tool for this is a large, sharp pocket knife. Cut down through the insulation to the copper conductor.

As you cut through the insulation DO NOT NICK the conductor. The trick is to avoid the perpendicular and to slice at an angle (Fig 3-2). Cut all the way around and pull off the insulation. The amount of protruding copper should be long enough for your purpose, not too long or too short.

If the copper end is too long, you might find that after installing it you would have to redo your work because it was preventing you from making a neat and safe connection.

Should the amount sticking out be too short, then you'll have to skin the copper a little more, again consuming time needlessly.

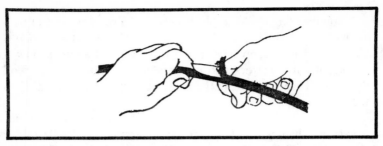

Fig. 3-2. Cutting the insulation at an oblique angle with a pocket knife.

If you nick the copper, you weaken it; and after a few unavoidable bends, it could possibly break where you nicked it.

Another problem a nick could cause is reduced ability of the wire to pass current. The heating of the wire during current flow would pose a hazard.

To eliminate the possibility of scoring the copper, use one of the many wire strippers available.

WIRE STRIPPERS

Skinning wire, to a wireman, is a fine art. The quick and error-free removal of insulation can readily be accomplished with special wire stripping tools (Fig 3-3).

One such tool is the flat wire stripper with wire-sized grooves in the jaws. The common version strips wire from size #18 to #10. (These are the wire sizes you'll be stripping most frequently.)

Using this tool you can't make an error. You simply insert the wire into the proper-sized groove in the jaws and squeeze the handles. The cutting edge of the stripper slices neatly through the insulation, down to but not touching the copper. A slight tug then slips the severed insulation from the copper.

Hold on. You are not finished yet. The insulation is cut straight through with the strippers. Good technique requires a slope in the

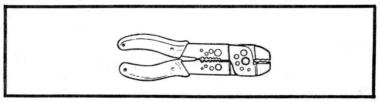

Fig. 3-3. There are any number of wire strippers available from the special to the combination cutter-stripper.

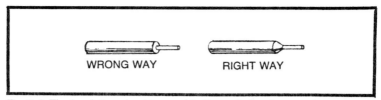

Fig. 3-4. The insulation should resemble the end of a sharpened pencil.

insulation as the illustration shows. You can whittle a slope with your sharp pocket knife in the way you sharpen a pencil. Then and only then, the wire end is ready for the connection (Fig. 3-4).

Get used to skinning wire carefully. There is only one right way to expose the wire end.

TERMINALS

There are, in general, two kinds of terminals to which you connect the wire ends. One terminal requires a loop at the end of the wire while the second allows you to insert the straight wire into it (Figs. 3-5 and 3-6).

The terminal requiring a loop is installed with a screw. The terminal is usually shaped to prevent the wire from getting away as you tighten the screw.

The screw itself comes with a lock bottom. Thus the screw cannot become detached and lost.

All terminal screws have right-handed threads.

When you form your loop in the wire end, don't make it too far or too near the margin of the insulation. The loop should consist of a single full bend.

Put the bend under the screwhead in a clockwise fashion, so that when you tighten the screw, the wire simply gets tighter. If you

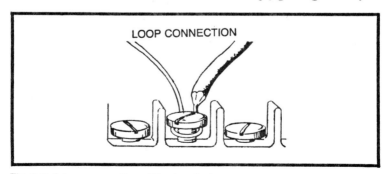

LOOP CONNECTION

Fig. 3-5. A loop connection at the terminal.

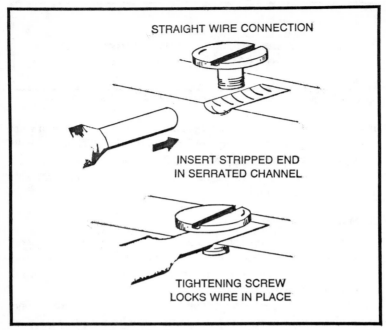

STRAIGHT WIRE CONNECTION

INSERT STRIPPED END
IN SERRATED CHANNEL

TIGHTENING SCREW
LOCKS WIRE IN PLACE

Fig. 3-6. A straight wire connection at the terminal.

had placed the bend in a counterclockwise position, the wire would have loosened as you tightened the screw (Fig 3-7).

But before you tighten the screw, trim the tip of the wire so that it fits around the screw as snugly as possible. Take care and do it exactly. If the connection is faulty, you're asking for trouble.

For tightening the screw down on the wire, choose the right screwdriver. In fact, since many of the terminal screws you'll be working with will have slots of about the same size, get yourself a screwdriver or two with blades closely suited to the task. This will insure you of getting a good, tight connection.

STRAIGHT WIRE CONNECTIONS

In recent years, because of the extra time a looped connection takes, straight wire connections have become popular. These have no regular terminal screw. A look at the terminal shows a hole and a set screw protruding into the hole. When you insert the wire, only a quick movement is required to tighten the set screw.

Another kind of terminal connector device does have a screw, but you do not have to loop the wire end around it. You'll find a serrated channel under the screw. The straight wire is inserted into the channel and the screw is tightened down. A good pressure connection is made.

Still another kind of terminal has no screws at all, neither terminal nor set types. You push the wire into a hole and a good connection is made automatically by a spring mechanism. To remove the wire, press a small screwdriver into a release slot.

LEARNING THEORY

Your mind is a wondrous tool that miraculously solves problems and plans procedures in electrical wiring. To accomplish these things, you must fill your mind with the bits of theory that are part of the electrician's craft.

Once the information is settled permanently in your mind, as you do a job the knowledge will take over and automatically direct your fingers to grasp the correct tools and apply them at the correct places till the wiring job is complete.

The theory becomes electrical wiring thinking patterns. The thinking patterns are formed from your ability to visualize the finished wiring job. Once visualized, then you set about to calculate the quantities involved.

You will calculate with formulas made up of fundamentals. The *electrical fundamentals* are numerous. The first ones you'll work with are volts, amperes and watts. Other fundamentals are ohms, inductance, capacity, reactance, impedance and phase. Then there are power factors, volt-amperes, gausses and frequency.

You might already be familiar with many of the physical fundamentals. You understand weight, volume, surface area and heat. These characteristics are expressed in standard U.S. units and in metric units. The electrician needs to work with all of these fundamental measurements.

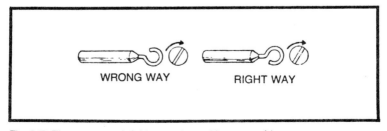

WRONG WAY   RIGHT WAY

Fig. 3-7. The wrong and right ways to position an end-loop.

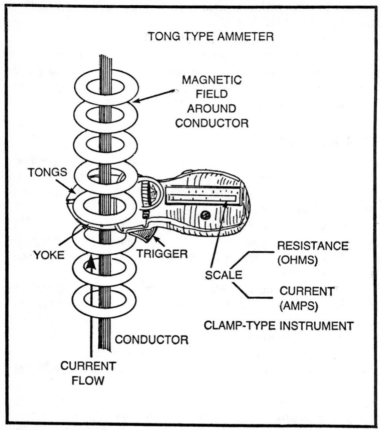

Fig. 3-8. Amperes are measured with a tong type meter by opening the tongs and encircling the wire carrying the current. (Courtesy of Amprobe Instrument)

METERS AND INSTRUMENTS

The electrician often has a need to measure voltage, amperage, resistance and watts. If he can measure these four quantities, he can accomplish most jobs.

There are many instruments for measuring these quantities. Most familiar is the tong type meter. It gets the name because a set of tongs at one end opens and closes when the meter is squeezed.

The tongs are there to measure amperage, which is the rate of flow of the current in the wires. To measure the amperage, all you do is open the tongs and encircle the wire, without actually making contact (Fig. 3-8).

The current flow is producing a magnetic field around the wire. The meter measures the amount of electromagnetism and presents a reading on the meter face.

To measure voltage, the tongs are ignored and two probes are used. If you'd like to measure the voltage in an outlet, you insert the probes into the two holes of the receptacle. (Forget the third hole—it's a ground.) The voltage pressures the meter and the needle on the meter face indicates the amount of voltage on the meter face.

To measure resistance, the same two probes are used, but first the current is switched off. You cannot measure resistance directly in an electrical system which is live. The power must be off.

Once the system is off, you can switch the meter to the resistance function and measure the resistance from point to point. The meter has a battery which sends a tiny amount of current into the system being tested. The battery's current is sufficient to move the meter needle and give a resistance reading.

This meter is a VOM type. The letters stand for Volt Ohm Meter. The instrument cannot measure watts directly. To do that you'll need a watt meter, a device that measures watts directly.

HANDY TESTERS

While scale-indicating meters are very helpful, electricians on the job use a handier type of tester. A neon bulb attached to a couple of test leads (Fig. 3-9) tells the electrician if a circuit is live right

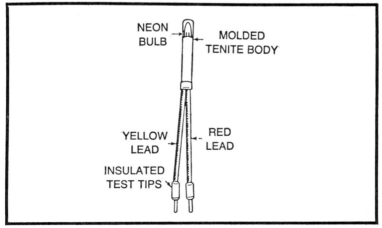

Fig. 3-9. The commonest test an electrician performs is to find out if a circuit is live. Simple testers show this. (Courtesy of Littlefuse, Inc.)

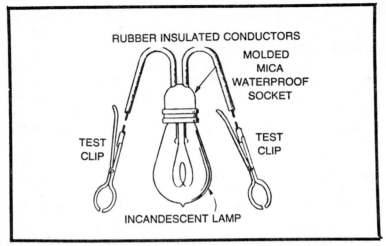

RUBBER INSULATED CONDUCTORS

MOLDED
MICA
WATERPROOF
SOCKET

TEST
CLIP

TEST
CLIP

INCANDESCENT LAMP

Fig. 3-10. This tester uses clips and an incandescent bulb.

quick. This is really all he wants to know about voltage and current most of the time anyway.

The neon bulb lights up in response to the live circuit. Whether a circuit is 117 volts or 120 volts is not of much interest in many situations. More often it's only of importance to know if a circuit is 120 volts or 240 volts. The neon bulb will shine twice as bright when applied to a 240-volt circuit. With a little practice you'll be able to tell how much voltage is in a line by the brightness of the bulb.

Resistance of a conductor or circuit is usually not important as far as actual measurement goes. What you usually need to know is whether there is "continuity." Is the circuit open or is it closed?

Electricians use a continuity tester. If the circuit is closed or shorted, the continuity is complete. The tester will be fitted with either a bulb that lights up (Fig. 3-10) or a doorbell that rings.

ELECTRICAL DRAWINGS

Vital to the electrician's trade are electrical drawings. These drawings are careful, precise plans of the job the electrician is to perform. Most often the electrical symbol instructions are presented on a set of "prints."

Print is a contraction of the word "blueprint." Several decades ago work drawings were printed on blue paper with white lines and symbols. Today most prints are on white paper with black or brown lines. Whatever the method for displaying them, the plans of the designing architect should meet all necessary specifications and standards. The print is the most compelling of the informational materials submitted to authorities when a building permit is sought.

On the print appear all of the details for the entire construction. Included are the plumbing, structure millwork, interior details, etc., along with the electrical symbols and lines. The electrician, though, only has eyes for the electrical information. The electrical symbols leap out at him as he reads the print (Fig. 3-11).

The symbols and initials have great meaning to him. The symbols designate general locations of equipment and wiring. As he examines the print, he will pinpoint precise locations of the wiring components and he'll gain a good idea of the schematic diagram of the system (Fig. 3-12).

The schematic diagram of the job, when it is available, is more detailed than the architect's rendering (Fig. 3-13). The schematic shows the actual circuit wiring and connections.

The symbols and lines of the schematic prescribe exactly how the system will be laid out, connection by connection. The architect's rendering does not show these details.

Many manufacturers and other electrical concerns produce a super schematic. Instead of schematic symbols, a pictorial rendition (Fig. 3-14) of the system and all its components is provided. This removes all of the guesswork. One can see how to connect the conductors to the components, without the extra mental step of relating a symbol to a device.

USING AN ARCHITECT'S SCALE

You should include an architect's scale in your tool box. It's a three-sided ruler that aids you in reading the print. The architect used one when he made the electrical drawings.

On the print is a listing like: ⅛ in. = 1 ft. This means that every ⅛ inch on the drawing represents 1 foot in the actual system. If a wall measures 2 inches on the drawing, it will be 16 feet high in the actual building. There are sixteen ⅛-inch segments in 2 inches.

Typical scales use dimensions of ⅛", ¼", ⅜", ½", ¾" or 1". Each equals 1 foot. For a house with a floor area of about 1500 square feet, a scale of ¼" = 1' is convenient. For a larger building, ⅛" = 1' might be better, to keep the print to a reasonable size.

Often the plans are drawn up by an engineer rather than an architect. Engineers also use a three-sided scale, but it's not the same as the architect's. The engineer's scale divides the foot into

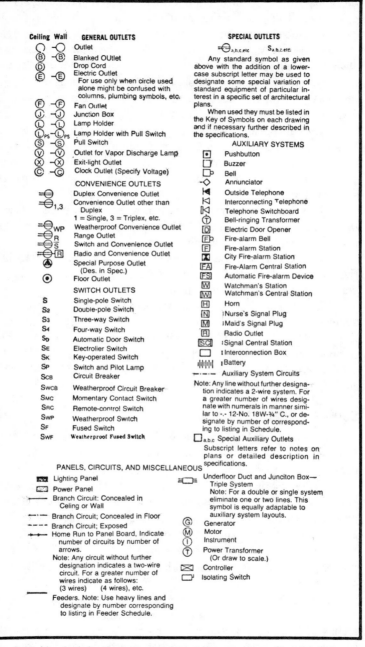

Ceiling Wall

GENERAL OUTLETS

Outlet
Blanked Outlet
Drop Cord
Electric Outlet
For use only when circle used alone might be confused with columns, plumbing symbols, etc.
Fan Outlet
Junction Box
Lamp Holder
Lamp Holder with Pull Switch
Pull Switch
Outlet for Vapor Discharge Lamp
Exit-light Outlet
Clock Outlet (Specify Voltage)

CONVENIENCE OUTLETS

Duplex Convenience Outlet
Convenience Outlet other than Duplex
1 = Single, 3 = Triplex, etc.
Weatherproof Convenience Outlet
Range Outlet
Switch and Convenience Outlet
Radio and Convenience Outlet
Special Purpose Outlet (Des. in Spec.)
Floor Outlet

SWITCH OUTLETS

S   Single-pole Switch
S₂   Double-pole Switch
S₃   Three-way Switch
S₄   Four-way Switch
S₀   Automatic Door Switch
Sₑ   Electrolier Switch
Sₖ   Key-operated Switch
Sₚ   Switch and Pilot Lamp
S_CB   Circuit Breaker
S_WCB   Weatherproof Circuit Breaker
S_MC   Momentary Contact Switch
S_RC   Remote-control Switch
S_WP   Weatherproof Switch
S_F   Fused Switch
S_WF   Weatherproof Fused Switch

SPECIAL OUTLETS

$=\bigominus_{a,b,c, etc.}$ S_{a,b,c, etc.}

Any standard symbol as given above with the addition of a lower-case subscript letter may be used to designate some special variation of standard equipment of particular interest in a specific set of architectural plans.

When used they must be listed in the Key of Symbols on each drawing and if necessary further described in the specifications.

AUXILIARY SYSTEMS

Pushbutton
Buzzer
Bell
Annunciator
Outside Telephone
Interconnecting Telephone
Telephone Switchboard
Bell-ringing Transformer
Electric Door Opener
Fire-alarm Bell
Fire-alarm Station
City Fire-alarm Station
Fire-Alarm Central Station
Automatic Fire-alarm Device
Watchman's Station
Watchman's Central Station
Horn
Nurse's Signal Plug
Maid's Signal Plug
Radio Outlet
Signal Central Station
Interconnection Box
Battery
Auxiliary System Circuits

Note: Any line without further designation indicates a 2-wire system. For a greater number of wires designate with numerals in manner similar to -.- 12-No. 18W-¾" C., or designate by number of corresponding to listing in Schedule.

Special Auxiliary Outlets
Subscript letters refer to notes on plans or detailed description in specifications.

PANELS, CIRCUITS, AND MISCELLANEOUS

Lighting Panel
Power Panel
Branch Circuit: Concealed in Ceiling or Wall
Branch Circuit; Concealed in Floor
Branch Circuit; Exposed
Home Run to Panel Board, Indicate number of circuits by number of arrows.
Note: Any circuit without further designation indicates a two-wire circuit. For a greater number of wires indicate as follows: (3 wires) (4 wires), etc.
Feeders. Note: Use heavy lines and designate by number corresponding to listing in Feeder Schedule.

Underfloor Duct and Junction Box—Triple System
Note: For a double or single system eliminate one or two lines. This symbol is equally adaptable to auxiliary system layouts.
Generator
Motor
Instrument
Power Transformer (Or draw to scale.)
Controller
Isolating Switch

Fig. 3-11. You look for these types of symbols as you read a print. They are among all the other construction details.

tenths and multiples of ten; the architect's scale divides the foot into inches (twelfths) and the inches into sixteenths. You'll probably have to own two scales, one for the architect's old fashioned renderings and one for the engineer's space age drawings.

The scales enable you to read locations of fixtures and receptacles, to plan sites for appliances, determine lengths of wire runs, and decide whether certain equipment will fit into tight places, and so on.

ELECTRICIAN'S USE OF PLOT PLAN

The plot plan is part of the set of prints. It has extensive information on it, like a legal description of the property, the configuration of the building and grounds, the setback distances of all structures, and much more. Little of this information is of major concern to the electrician.

What is of concern is where the power company has its poles and wires, or whether the power company's lines are underground.

This allows you to estimate within inches the *point of entry* of the electrical service. It tells you the height you will install your *service head* and you can figure the Code-required clearances of the power company's *service drop* to the building.

Often some of the information you need is not on the print. Then you must get together with the power company's wire people and work out certain design points. These points should then be drawn to scale on the plot plan. After the power company people have verified your drawings, you are all set for the actual installation.

The plot plan can also be used to mark the path of the underground wiring from building to building, and to swimming pools, post lights, landscape and gate lighting, etc.

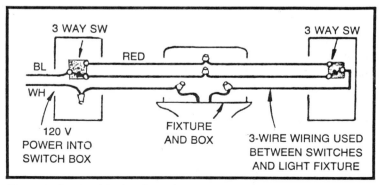

Fig. 3-12. An electrician's schematic diagram.

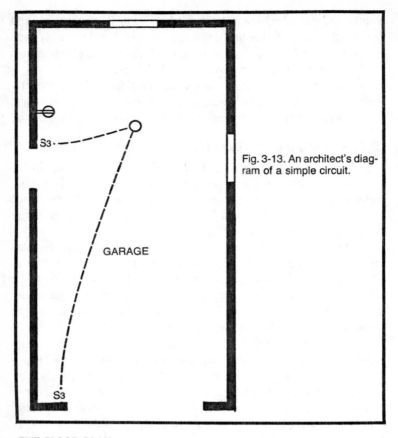

Fig. 3-13. An architect's diagram of a simple circuit.

THE FLOOR PLAN

The highly important instructions from the architect or engineer to you appear on the floor plan. From your point of view, a good floor plan is one which is largely an electrical floor plan. The floor plan shows the locations of the doors, windows, closets, bathroom and kitchen fixtures, partitions and so on where you are going to install the electrical components.

Beneath the area drawing on an electrical floor plan appears a list of symbols. These represent all the outlets, switches and special items the home designer wants installed.

For example, when you consult the floor plan, you can see on circuit number 4 a dashed line connecting wall outlets, ceiling outlets, a 3-way switch, a single-pole switch, two duplex convenience outlets on either side of a partition, two duplex outlets and one weatherproof convenience outlet on either side of an outside wall, and an outside wall outlet.

The point of entry for the service to the meter is shown on the basement floor plan. Near the meter are the distribution panel and six circuits. Circuit number 6 doesn't go anywhere since it is the spare circuit.

Circuit number 5 is shown in the basement and then again leading upward near the stairway. It reappears on the first floor at the head of the stairwell. It goes to a group of switches, then to the kitchen ceiling outlet, into a clock outlet, a duplex convenience outlet, and terminates at the electric range.

Circuits 1, 2, and 3 can be followed the same way. It might seem sketchy, but actually there is enough information for installing an electrical system. With this information an entire electrical system can be installed. This means purchasing only approved materials, installing them in a professional way, and having the completed job pass all necessary inspections.

A REAL JOB

Getting down to actual facts, if the electrician is handed a residential wiring job, the print is only a guide. Yet it enables him to actually lay out his work after he has referred to it to determine certain essential information. He will use the print to figure out the square footage of the home. (Garages and open porches are not

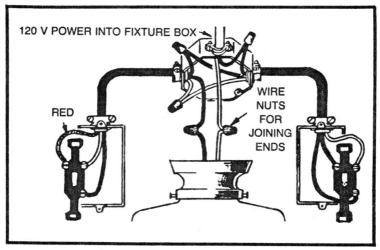

Fig. 3-14. A pictorial rendition of a lamp-and-switches arrangement.

included in the calculation.) And he will use the print to find out the number and kinds of heavy duty appliances to be used in the home.

A typical group of heavy duty appliances would be composed of a dishwasher, garbage disposer, range, trash compactor, water heater, clothes dryer, air conditioner, electric space heater, and perhaps a motor or two (for a pump and swimming pool). Smaller wattage appliances like washing machines, electric irons, and TVs are not considered to be heavy duty appliances.

Once he obtains the above information, the electrician can calculate the kilowatt load of the entire system. This is an important quantity, because with it he can figure the size of his wires, the size of his service, the size of his distribution panel, and the size of his circuits, bends and other things.

When he has an actual layout, he can go to work. If he has learned how to calculate the wiring design, he can consider himself capable of advanced electrical work. This kind of calculating will be covered in Chapter 14.

PRINTS WITH HARDLY ANY ELECTRICAL DATA

On many jobs—especially on repair and or maintenance jobs—the wireman may be given little basic information. Even his instructions will be scanty.

On a residential job the homeowner or builder might merely say, "Install a system to carry the following load." So the wireman looks over the load.

The house has 1800 square feet of living area and contains a 1.5-kW cooktop, two 3-kW ovens, a 4-kW water heater, a 6-kW air conditioner, an electric dryer, and a 10-kW electric heater in each of the five rooms.

That's it. That is all the information given, and actually that's all that's needed. As you make your way through this book, you will learn all the fill-in information to enable you to carry out a job like this.

That is what electrical wiring is all about: the know-how that takes up only a few sentences but equips you to conceive and install a complete electrical system.

Ohm's Law and the Power Formula

Four distinguished scientists of the early 1800's had electrical characteristics named in their honor. Georg Simon OHM found that the ability of electric current to pass through various materials depended upon the nature of the material. Certain materials passed the current easily, others not. Each material offered its own degree of RESISTANCE.

Silver was found to CONDUCT almost all of the electricity applied to it, while copper was only slightly less efficient.

On the other hand, rubber was found to pass practically no current and, in effect, was acting to INSULATE the electricity. Most other materials, whether metal or nonmetal, were found to conduct electricity better than rubber but not as well as copper. The inherent resistance of a material to the flow of electricity, as you know, is measured in *ohms*.

The other three scientists were James WATT, Alessandro VOLTA and Andre AMPERE. All four gentlemen, due to their investigations into fundamental facts of nature, will have their names perpetuated in science books for all time.

AN ELECTRIC CIRCUIT

When you flick a wall switch and the overhead light goes on, you have closed it with the switching action.

An electric circuit is composed of four entities. First of all, there is the source of the current. The source of the current for the

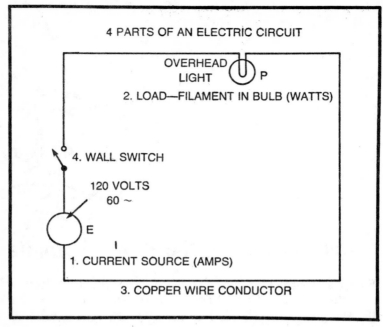

4 PARTS OF AN ELECTRIC CIRCUIT

OVERHEAD LIGHT P

2. LOAD—FILAMENT IN BULB (WATTS)

4. WALL SWITCH

120 VOLTS
60 ~

E

I

1. CURRENT SOURCE (AMPS)

3. COPPER WIRE CONDUCTOR

Fig. 4-1. This electric circuit is made up of four parts. The source, load, conductor and switch.

overhead light of the last paragraph is the electric company. The source is sending out 120 volts of electrical pressure (Fig. 4-1).

The electric company supplies the electric pressure, whether the light is turned on or not. Similarly, the water company supplies water pressure whether you have the faucet turned on or not.

The second entity of an electric circuit is the load. In the case of the overhead light, the filament of the bulb is the load.

The bulb has a certain wattage. The bulb consumes watts while it is turned on. It doesn't consume any watts while it is off. You've seen the electric company's meter. It is a wattmeter. It measures the number of watts used whenever current is flowing. The meter records the current flow in units of measurement called kilowatts. Kilowatt merely means one thousand watts. It would be tedious to measure watts one by one since such great quantities of them are being consumed constantly (Fig. 4-2).

The third entity of a circuit is the conductor. Copper wire is used commonly and aluminum wire occasionally. Both are satisfactory conductors. The copper conducts better than the aluminum, but the aluminum is cheaper. Both are able to conduct the current that is being pushed by the voltage. The current is measured in amperes. Specifically, the amperes express the current's rate of flow.

The fourth entity (perhaps easier thought of as a component) of a circuit is the switch. The switch is situated somewhere along the conductor line. Its sole function is to open and close the circuit.

OHM'S LAW

In electrical calculations you quite often use Ohm's Law, yet you do not use ohms in the figuring. Instead, you use a related Power Formula.

In the electric circuit just discussed, the ohm was not mentioned. Only volts, amps and watts were mentioned. Actually, in most calculations of everyday work, especially in the realm of residential wiring, the ohm is seldom used—at least directly. The ohm *does* underlie the calculations, however. If you care to dig deep, you'll be able to understand the reasons. But make no mistake, the ohm is a crucial consideration in every circuit.

Meanwhile, let's look at the relationship between volts, amps and watts, abbreviated E, I and P, respectively.

KILOWATT-HOUR METER

READ POINTERS FROM LEFT TO RIGHT

THIS READING IS 66,482 KILOWATT-HOURS

THIS READING IS 66,649 KILOWATT-HOURS

TOTAL BETWEEN TWO IS 167 KILOWATT-HOURS
WHICH IS QUANTITY UTILITY CO. BILLS YOU FOR

Fig. 4-2. The load when energized consumes power and is measured on the kilowatt-hour meter.

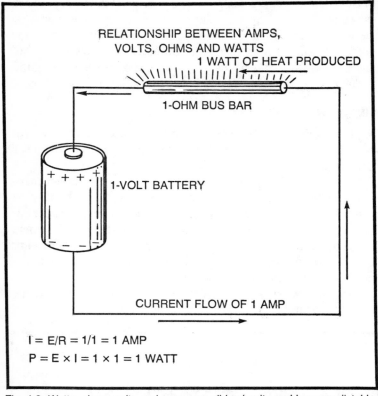

RELATIONSHIP BETWEEN AMPS,
VOLTS, OHMS AND WATTS
1 WATT OF HEAT PRODUCED

1-OHM BUS BAR

1-VOLT BATTERY

CURRENT FLOW OF 1 AMP

$I = E/R = 1/1 = 1$ AMP
$P = E \times I = 1 \times 1 = 1$ WATT

Fig. 4-3. Watts, ohms, volts and amps are all in circuits and have predictable relationships.

To standardize calculating procedures, the electrical pioneers declared that one watt of power is produced when one volt pushes one amp through a conductor. As a result, the Power Formula, $P = EI$, was born.

From simple algebra, if we juggle the formula, then $I = P/E$ and $E = P/I$. You must memorize these three formulas. You'll be using them constantly.

For example, suppose the voltage in our circuit is 120 and the bulb is rated at 100 watts. How much current does the bulb draw from the electric company?

$$I = P/E = \frac{100 \text{ watts}}{120 \text{ volts}} = 0.83 \text{ amps}$$

OHM'S LAW HAS RESISTANCE(R)

I'm sure you've all heard that Ohm's Law is really thought of as I = E/R. That formula was the original version of Ohm's Law. It is used in many electrical calculations, although in residential wiring not as often as P = EI.

I = E/R is a transposition of R = E/I. A load is said to have a resistance of one ohm when a pressure of one volt pushes one amp through it.

Yes, one watt of heat is produced. That is why the four units are related (Fig. 4-3).

It is used like this. Getting back to our simple light circuit, figure out the resistance of the 100-watt bulb's filament (Fig. 4-4).

We know that the voltage is 120 and we've discovered that the current being drawn by the bulb is 0.83 amperes.

$$R = E/I = \frac{120 \text{ volts}}{0.83 \text{ amps}} = 144 \text{ ohms}$$

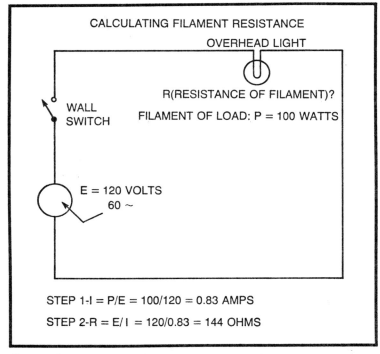

CALCULATING FILAMENT RESISTANCE

OVERHEAD LIGHT

R(RESISTANCE OF FILAMENT)?

FILAMENT OF LOAD: P = 100 WATTS

WALL SWITCH

E = 120 VOLTS
60 ~

STEP 1-I = P/E = 100/120 = 0.83 AMPS

STEP 2-R = E/ I = 120/0.83 = 144 OHMS

Fig. 4-4. Calculating the resistance of a simple circuit.

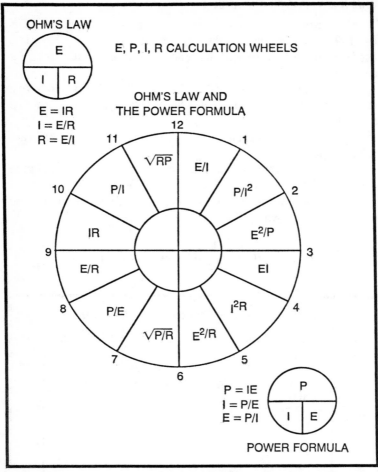

Fig. 4-5. The Ohm's Law Wheel is a convenient way to calculate circuit characteristics. Copy it into your Code Book.

Ohm's Law and the related Power Formula are vital to the electrician. He uses them constantly. Using them must become second nature to him. He must learn to manipulate the formula in a number of ways. One handy method for remembering the various ways in which Ohm's Law will work out is with the Ohm's Law Wheel (Fig. 4-5).

When you look at the wheel for the first time, confusion reigns. But if you look carefully, things start to make sense. The wheel has a hubcap in the center and 12 spokes.

The hubcap contains the four vital letters of the formual. R leads into spokes 1, 2 and 3; P goes to 4, 5 and 6; I to 7, 8 and 9; and E into 10, 11 and 12.

If you want to find R, you have the choice of three formulas, in spokes 1, 2 and 3. Each letter has three convenient formulas.

There is no need to memorize the wheel at first. Put it on a card and keep it in your tool box. Many electricians place a diagram of the wheel in their Code book for reference.

WHAT IS ELECTRICITY?

We have just discussed that volts push amps through resistance and cause wattage consumption. We've even learned how to calculate the various electrical characteristics. However, what exactly is electricity?

From an electrician's point of view, electricity resides in everything. It is in copper, coal and rubber. Electricity is electrons. All matter contains electrons.

If you were able to reduce yourself to the size of an electron and to enter inside a piece of copper wire, you'd find youself floating in empty space. The space is quite like the space between the earth and the sun. You'd see you were in a universe surrounded by billions of other universes. The electrons in the copper are revolving about a nucleus just as the earth and planets revolve about our sun.

The electrons are pieces of electricity and are locked into orbit around their own nucleus (Fig. 4-6). Each copper atom, composed of a nucleus and a number of electrons, is its own universe. A piece of copper wire is composed of billions upon billions of atoms.

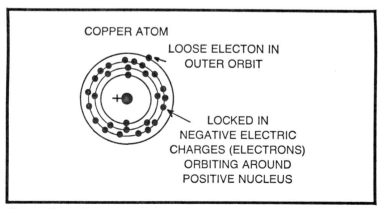

Fig. 4-6. Matter is composed of electrons in orbit around a nucleus.

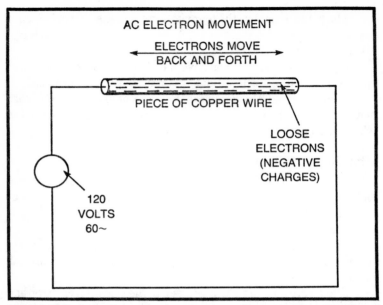

Fig. 4-7. As the loose electrons are repelled in a conductor, they tend to heat the conductor and produce a magnetic field around the conductor. This is the work that electricity performs.

A more detailed look into the copper wire will show that in addition to the copper atoms tightly locked in orbit, there are a great many loose electrons—bouncing from atom to atom in random fashion. You can see that with a bit of effort you could cause the loose electrons to move as you direct. You can realize also that it would be difficult to remove the orbiting electrons from their tight lock to the nucleus.

MAKING ELECTRICITY WORK

Each electron is considered to have a negative charge. Don't confuse the way the term negative is used here with the way it is used in numbers. The electron could have just as sensibly been called a positive charge, blue charge, wet charge, or what have you. It was called negative, however, and the name stuck. That is why ordinary electricity is known as negative electricity: because the pieces of electricity just happened to have been called negative.

Now, if you bring two electrons close together, they repel each other. Your inference must be that like charges repel each other. Thus, two electrically negative particles push each other away.

If you take a voltage source and attach it to a piece of copper wire, electrons will leave the source and flow into the copper wire (Fig. 4-7). According to the amount of electrical pressure (voltage), a lot or only a few electrons will flow into the wire.

The electrons flowing out of the voltage source will push along all the loose electrons in the copper. The orbiting electrons will be affected, too, but they will stay locked in orbit—unless the voltage becomes extraordinarily large.

The loose electrons will flee through the wire, repelled by the oncoming electrons. The passage of the loose electrons produces electrical work.

HEAT

As the electrons move, they create two effects. Number one is heat; as the loose electrons flow, they raise the temperature of the copper. Loads are designed to produce heat.

If the load is the filament of a bulb, it will get white hot and throw off light. If the load is the element of a clothes dryer, heat will be generated in the interior of the dryer and moisture will be evaporated from the clothes. If the load is the element of a space heater, the temperature of a room will be raised.

The movement of the electrons can be measured in amperes. A meter can be placed at any point in the circuit to measure this movement. It might be interesting to note that one ampere represents a flow of 6,280,000,000,000,000,000 electrons per second past a given point in the circuit. As mentioned, the rate of flow is referred to as I in the formulas.

If you would like to know the wattage of the heat produced by the current in a piece of wire, use Ohm's Law. On the wheel, $P = EI$, $P = I^2R$ and $P = E^2R$ on spokes 4, 5 and 6, respectively.

For example, suppose the current is 0.83 amps in a 120-volt circuit (Fig. 4-8). $P = EI$ or $120 \times 0.83 = 99.6$ watts. Or, suppose the resistance is given as 144 ohms. Then, $P = I^2R$ or $(0.83)^2 \times 144 = 99.2$. (Use 0.83333 and it will come out in both formulas to 99.9999.) Yes, this is the same overhead light circuit we've been discussing.

MAGNETIC FIELDS

The second effect that is created as the voltage impels the loose electrons along in an orderly fashion is the formation of a magentic field (Fig. 4-9).

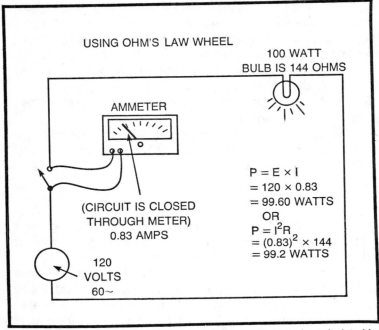

Fig. 4-8. The amount of wattage produced by a current can be calculated by using the Ohm's Law Wheel.

Whether you want to or not, you get a magnetic field around a wire whenever the electrons move. It can be useful or it can be a nuisance, but there is no way to eliminate it.

If you take a straight piece of wire and run some current through it, a magnetic field is instantly produced around the wire. It is an invisible force field that is strongest close to the wire.

Should the current suddenly stop, the force field collapses just as suddenly. When the current starts up again, the field snaps back into place.

As the current is flowing, the magnetic field rotates about the wire in a certain direction. When the current is reversed, the magnetic field reverses the direction of its rotation.

Loads are designed to use the magnetic field. If the load is the winding of a motor, the magnetic field causes the motor to start rotating. When the load is a solenoid, a core will be caused to move. Should the load be a relay, a movement of the relay occurs as the magnetic field is produced. In a transformer, alternating current is caused to transfer from a primary to secondary.

The magnetic field can be wasteful, too. With a number of conductors in a conduit, the magnetic field can heat up the conduit and consume wattage to no advantage.

While we are are discussing magnetic effects, you can quickly learn the electrician's Left Hand Rule. With it you can tell the direction in which the magnetic field rotates.

If, while direct current is flowing in a wire, you wrap the fingers of your left hand around the wire with your thumb in the direction of the electron flow, your fingers will point in the direction of the magnetic field's rotating lines of force.

PRODUCING ELECTRIC CURRENT

Not only does a current produce a magentic field, a magentic field can produce a current. If you move any permanent magent (bar, horseshoe or whatever) past a wire, the loose electrons of the wire will be induced to move.

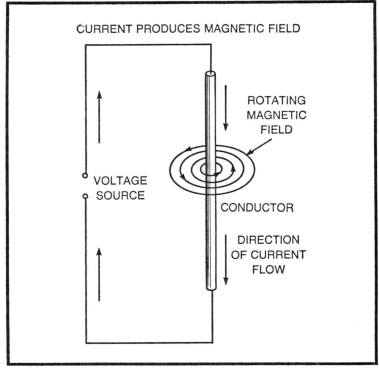

Fig. 4-9. The flow of electrical current creates a rotating magnetic field around the conductor.

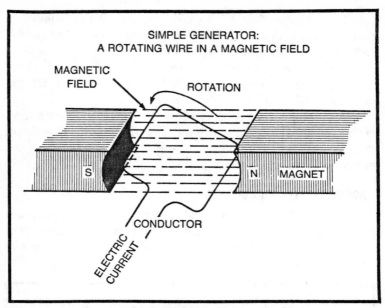

MAGNETIC
FIELD

ROTATION

S

N   MAGNET

CONDUCTOR

ELECTRIC
CURRENT

Fig. 4-10. A simple generator can be made by rotating a coil of wire in a magnetic field. As the coil rotates, current flows back and forth.

This constitutes an electric current. If you suddenly stop moving the magent, the electrons will stop flowing. As you start moving the magnet again, the electrons will begin flowing.

The electrons move in a definite direction every time. They will move as if they were attached to the magnet. Actually they are attached to the force field surrounding the magent.

A magnet has two poles, sometimes referred to as north and south poles. One pole attracts the electrons, while the other pole repels the electrons. The magnet holds the electrons in its power. When you move the north pole near the wire, the electrons flow in a certain direction. Should you then rotate the magnet so that the south pole occupies the position previously held by the north pole, the electrons will reverse the direction of their flow.

If you affix the magnet to an axle and spin the axle so that the north and south poles rotate alternately toward and away from the wire, the electrons will flow one way and then the other. You have generated an alternating current (AC).

Since electrical energy consists of nothing more than the flow of electrons in a conductor, the simple arrangement of the last paragraph is in fact an AC generator. Commercial AC generators are designed to operate under an arrangement whereby a conductor is rotated within a magnetic field (Fig. 4-10).

INSULATORS

You'll recall that the copper wire not only contained a host of electrons locked in orbit around the nuclei of atoms, it also contained an abundance of loose electrons bouncing around randomly among atoms.

A piece of glass is an insulator (the opposite of a conductor), and the loose electron situation inside is not the same. The piece of glass has plenty of atoms with electrons in orbit, to be sure; and the electrons behave no differently than those of copper atoms. After all, an electron is an electron. However, you cannot find any loose electrons. All of the glass's electrons are locked tightly in orbit (Fig. 4-11).

It works out that in every atom of glass the sum of the negative charges (each orbiting electron represents one negative charge) exactly equals the sum of the positive charges in the nucleus. It is difficult and requires great amounts of voltage to knock the orbiting electrons out of their mating position with the nucleus. Under normal voltage conditions, the glass will not pass any electrons and thus there will be no electric current.

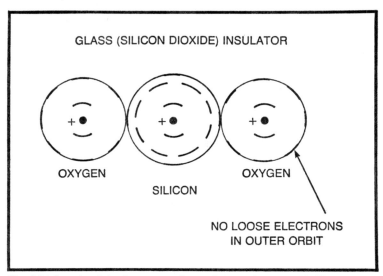

Fig. 4-11. In an insulator there are no loose electrons in outer orbit. Electrons are locked in tightly.

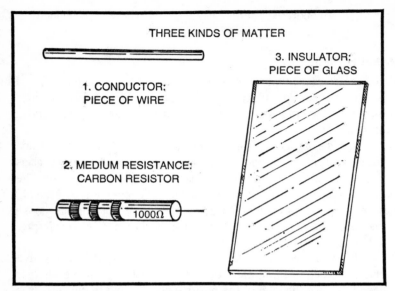

Fig. 4-12. A carbon resistor is neither an outstanding conductor nor an outstanding insulator. It has an average resistance.

That is how an insulator works. When insulation is wrapped around a piece of copper conductor, you can hold the wire and be safe from the current. The electricity cannot pass through the insulation, normally.

An important test is the measurement of an insulator's resistance. While a conductor may only have a resistance of a fraction of an ohm in a span of a hundred feet, an insulator can easily have a resistance of 500,000 ohms in 1/64 of an inch!

RESISTANCES

Between conductors with very low ohmages and insulators with extremely high ohmages, there exist materials with average resistances (Fig. 4-12). Normally, though, when we speak of resistances, we are referring to the circuit's load.

We've seen that a 100-watt light bulb has a resistance of 144 ohms. We can easily calculate the resistances of a 50-watt bulb and a 500-watt bulb in a 120-volt circuit.

$$R = E^2/P \text{ (spoke \#3 of the wheel)}$$

$$R = \frac{(120)^2}{50 \text{ watts}} = 288 \text{ ohms}$$

$$R = \frac{(120)^2}{500 \text{ watts}} = 28.8 \text{ ohms}$$

Let's calculate the resistance of a 5-kilowatt electric dryer in a 240-volt circuit (Fig. 4-13).

$$R = E^2/P = \frac{(240)^2}{5000 \text{ watts}} = 11.52 \text{ ohms}$$

It is apparent that certain loads could be so slight that their resistance would scarcely exceed that of the conductor. It is also apparent that small loads like the 50-watt light bulb have resistances that are negligible when compared with the resistance of the conductor's insulation.

You can also see that as the load rises in wattage, it lowers in resistance. As the wattage of the load declines, the resistance rises.

The characteristics of the relation of resistances to loads should be ever on the mind of the electrician.

PRACTICE WITH OHM'S LAW

Let's get back to our overhead light circuit. It is a good example of the way an electrician should think about Ohm's Law.

If you look at Fig. 4-14, you can see that the circuit conductor consists of two 50-foot pieces of #14 wire. From the 15-amp circuit

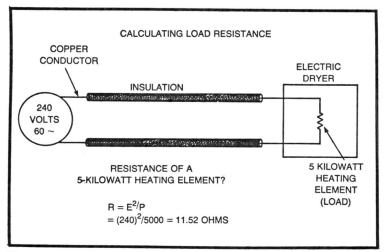

Fig. 4-13. The resistance of the heating element of an electric dryer can be calculated quickly.

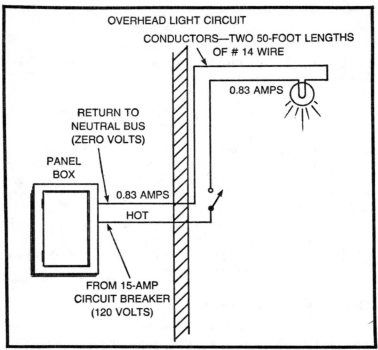

Fig. 4-14. Across a load, like the light bulb, the voltage drops from 120 volts to zero, but the current remains 0.83 amps throughout.

breaker in the panel box, a wire rises through the walls and connects to the ceiling light. This is the hot wire and it will measure 120 volts if tested by a meter.

A second wire leaves the light fixture, travels across the ceiling, descends inside the wall and connects to the neutral bus in the panel box. If you measure the voltage on this neutral wire, the meter will read zero.

The voltage, then, ranges from 120 volts on the hot side, to zero volts on the neutral side. However, the current, which we have calculated to be 0.833 amps, is the same throughout the circuit. The current is 0.833 amps on the 120-volt hot side and 0.833 on the zero-volt neutral side. Remember, the current is the rate of flow of electrons; the voltage is only the pressure on the electrons.

VOLTAGE DROP ACROSS LOAD

Where precisely does the voltage drop from 120 volts to zero volts? Does it drop across the switch, across the 50-foot piece of

#14 hot wire, across the 50-foot piece of #14 neutral wire, or does it drop across the load?

It drops across the resistance of the load. The resistance of the filament in the light bulb is 144 ohms. As proof that the 120-volt drop occurs here, we can apply Ohm's Law.

Voltage across load: IR or 0.833 × 144 = 119.952. For all intents and purposes, the answer is 120 volts.

Notice that the voltage drop across the load is calculated by IR. That is why a voltage drop is also called an IR drop.

VOLTAGE DROP ACROSS CONDUCTORS

That last example wasn't too difficult to envision. However, voltage drops can sometimes be puzzling. Let's explore our overhead light circuit further.

Suppose a short circuit develops in the bulb socket and shorts the current across the gap between the hot conductor and the neutral conductor. This is possible because the two wires lie close together at the terminal connection. A short circuit could happen in a number of ways. A path could be burnt in the insulator separating the two terminals; a piece of wire could accidentally fall across the terminals; or other problems could occur. When they do, blooey!

What happens according to Ohm's Law? In effect, the two 50-foot pieces of #14 wire become the load. The 120 volts get dropped through the 100 feet of wire.

What happens to the current? I = E/R will tell us. R can be obtained from Table 8 (Chapter 9) in the NE Code book. One hundred feet of #14 wire has a resistance of 0.257 ohms.

$$I = \frac{120}{0.257} = 466.93 \text{ amps}$$

If you consider that the entire service in a large home is 200 amps and that this circuit was drawing only 0.833 amps, then you appreciate what this short circuit is drawing in current—466.93 amps!

What really happens, though, is that immediately the short occurs and the dangerously large current begins to flow, the 15-amp circuit breaker opens and isolates the circuit. There is no danger in a properly designed system.

VOLTAGE DROP ACROSS CONDUCTORS AND LOAD

An even more puzzling question about voltage drop arises when you seek to learn how much voltage is dropped across the conductors under normal load conditions.

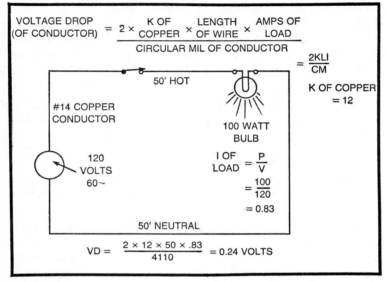

Fig. 4-15. To calculate the voltage drop in the conductors, this formula is used. The K is the constant of the material. L is the length of wire one way. The 2 multiplies the length of wire so the entire length is accounted for. I is the amps of the load. CM is the circular mils of the conductor taken from Chapter 9, Table 8 of the Code book.

A little earlier we mentioned that the load drops the full 120 volts. The statement is not exactly true. The load drops most of the 120 volts, but not all of it. The 100 feet of conductor is responsible for a small fraction of the drop.

Bear in mind that the current doesn't vary in this type of circuit. Its value is the same in every part of the circuit, no matter the location of the voltage drop.

To determine the voltage drop of the conductors, you use a more complicated formula than Ohm's Law. The formula for voltage drop is derived from Ohm's Law but is modified to account for other considerations.

What you should remember is that although the load is calculated using Ohm's Law, the voltage drop of the conductors in the circuit is not. Your inclination is to regard the load as being responsible for the entire voltage drop.

The conductor voltage drop is calculated with the formula:

$$\text{voltage drop} = \frac{2 \cdot K \cdot L \cdot I}{CM}$$

In the above formula, K is a constant. (Use a value of 12 when K is copper and a value of 18 when K is aluminum.) L is the length of the conductor in one direction. (The hot wire extends in one direction; the neutral wire in the other.) I is the current of the load. CM is the circular mil area of the wire. This is also obtained from Table 8 (Chapter 9) of the Code.

If we assume that the conductor is of #14 copper, we can calculate the voltage drop in our overhead light (Fig. 4-15) by the following formula:

$$\text{voltage drop} = \frac{2 \cdot K \cdot L \cdot I}{CM}$$

$$= \frac{2 \cdot 12 \cdot 50 \cdot 0.833}{4110}$$

$$= \frac{996}{4110}$$

$$= 0.24 \text{ volts}$$

That is about ¼ volt—a value which is negligible in this circuit. The Code recommends a voltage drop of no more than 3%, or, 3.6 volts in a 120-volt circuit.

Ohm's Law is vital to the electrician. It is really not complicated but for correct use does require some visualization of the circuit.

Conduit

In the electrical trade, the word *conduit* refers to the tubular metal housing through which the current-carrying wires (the conductors) are strung in modern-day installations.

ADVANTAGES

Conduit, more commonly known as pipe, is considered to be an appropriate and normal material in today's electrical wiring systems. It is like a permament subway tunnel for the wiring, up and down, across ceilings, under floors, inside and outside (Fig. 5-1).

In systems installed with conduit you can always pull out old wiring, add new wiring and other such conveniences, and never have to open a wall or ceiling.

This is in constrast to a Romex installation, which is easy to work with while the walls are open, but not so easy after the walls have beeen closed (Fig. 5-2).

In addition to being esteemed for its wire changing capability, conduit is prized for the excellent, permanent path to ground it provides. This virtue of conduit enhances the safety of an electrical system. Shock hazard is at a minimum and fire protection at a maximum when conduit is used. Furthermore, because it is sturdy and smooth, conduit resists accidental puncturing.

When tucked safely inside conduit, electrical wiring is highly resistant to blows by hammers, being run over by wheelbarrows,

Fig. 5-1. Conduit forms a permanent tunnel for the wiring, up and down, across ceilings and all over.

the gnawing of small rodents, earth tremors, lightning strikes and so on.

The rigid type of metal conduit is built to resist the corrosive effects of cement and can be embedded without fear into slabs, partitions and walls.

EMT (Electrical Metallic Tubing), also called *thinwall* conduit, is not as rugged as rigid metal conduit, but affords the same kind of protection, though to a lesser degree. Figure 5-3 illustrates the two types of conduit.

ELECTRICAL METALLIC TUBING (EMT)

EMT is called thinwall, not because it will fit inside a narrow partition but because its wall thickness is less than that of rigid metal conduit. When the inside diameters of the two types of pipe are equal, EMT conduit will have the smaller outside diameter (Fig. 5-4).

Pipe is manufactured in a range of diameters from ½″ to 6″. These are called trade sizes (see Table 5-1, Column 1). The trade size is a rough index to the inside diameter of the pipe. Column 2 of Table 5-1 lists the actual inside diameter in decimal equivalents of

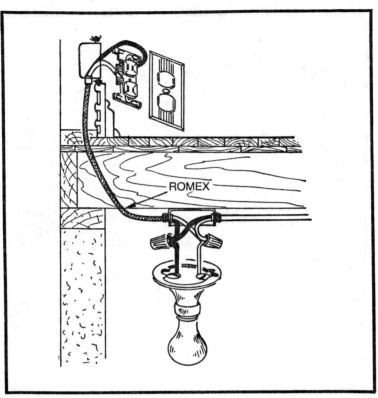

Fig. 5-2. A Romex installation, while just as adequate to carry current, is unchangeable once the walls are closed.

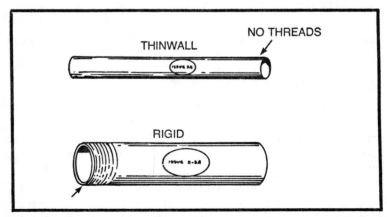

Fig. 5-3. These are two of the different types of pipe available for different kinds of installations.

inches. Note that the ½″ size has a diameter of .622 inches. The true value of ½, of course, is 0.500. You'll find that the actual pipe diameters are slightly larger than the trade sizes indicate.

The outside diameters of the pipe are not listed in the table. The outside diameters are available, but as a wireman you'll rarely, if ever, need to know them.

It is good to understand, though, that EMT has thinner walls than rigid metal conduit, and, since the inside diameters of both types are about the same, that EMT has smaller outside diameters. That is, up to the 2″ size. Beyond this point EMT has the same outside diameter as rigid metal conduit. This means that EMT inside

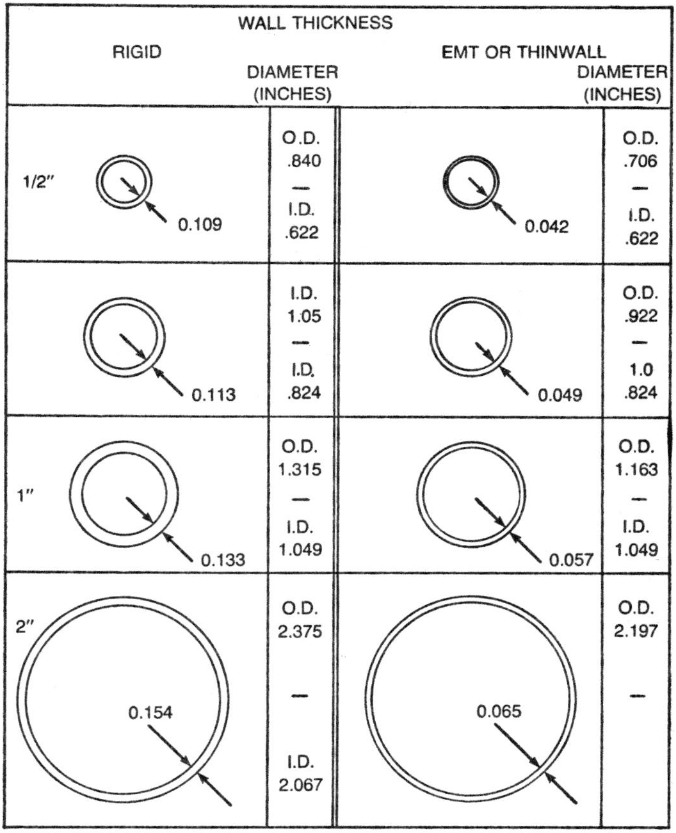

WALL THICKNESS				
RIGID		DIAMETER (INCHES)	EMT OR THINWALL	DIAMETER (INCHES)
1/2″ 0.109		O.D. .840 — I.D. .622	0.042	O.D. .706 — I.D. .622
0.113		I.D. 1.05 — I.D. .824	0.049	O.D. .922 — 1.0 .824
1″ 0.133		O.D. 1.315 — I.D. 1.049	0.057	O.D. 1.163 — I.D. 1.049
2″ 0.154		O.D. 2.375 — I.D. 2.067	0.065	O.D. 2.197 —

Fig. 5-4. As you can see, the inside diameters of rigid conduit and EMT are identical up to the 2″ size. Beyond this point the outside diameters are the same.

Table 5-1. Dimensions and Percent Area of Conduit and of Tubing (National Electrical Code)

Trade Size	Internal Diameter Inches	Total 100%	Nipples 60%	Not Lead Covered			Lead Covered				
				2 Cond. 31%	Over 2 Cond. 40%	1 Cond. 53%	1 Cond. 55%	2 Cond. 30%	3 Cond. 40%	4 Cond. 38%	Over 4 Cond. 35%
½	.622	.30	.18	.09	.12	.16	.17	.09	.12	.11	.11
¾	.824	.53	.32	.16	.21	.28	.29	.16	.21	.20	.19
1	1.049	.86	.32	.27	.34	.46	.47	.26	.34	.33	.30
1¼	1.380	1.50	.70	.47	.60	.80	.83	.45	.60	.57	.53
1½	1.610	2.04	1.2	.63	.82	1.08	1.12	.61	.82	.78	.71
2	2.067	3.36	2.02	1.04	1.34	1.78	1.85	1.01	1.34	1.28	1.18
2½	2.469	4.79	2.87	1.48	1.92	2.54	2.63	1.44	1.92	1.82	1.68
3	3.068	7.38	4.43	2.29	2.95	3.91	4.06	2.21	2.95	2.80	2.58
3½	3.548	9.90	5.94	3.07	3.96	5.25	5.44	2.97	3.96	3.76	3.47
4	4.026	12.72	7.63	3.94	5.09	6.74	7.00	3.82	5.09	4.83	4.45
4½	4.506	15.94	9.56	4.94	6.38	8.45	8.77	4.78	6.38	6.06	5.56
5	5.047	20.00	12	6.20	8.00	10.60	11.00	6.00	8.00	7.60	7.00
6	6.065	28.89	17.33	8.96	11.56	15.31	15.89	8.67	11.56	10.98	10.11

diameters (in sizes greater than 2″) are larger than those of rigid metal conduit.

To further complicate matters, the table shows average internal diameters of pipe. Different manufacturers produce pipe whose diameters differ slightly from standard diameters. Don't worry, just stick to the table for the calculations and you will be in compliance with Code rulings.

EMT INSTALLATION RULINGS

The Code allows you to use thinwall often, especially in residential and small commercial installations.

You can use EMT in both concealed and exposed situations. You can attach it to wood or to metal studs in walls. You can bury it in concrete or plaster or mount it on the face of masonry walls.

When a run of EMT has to pass through a cinder bed or an earthen fill, it must be protected on all sides by a minimum of 2 inches of concrete unless the run is laid more than 18 inches under the fill.

When an EMT installation is used in wet areas, like the laundry room, it has to be made waterproof. Water must not be allowed to enter the pipes. The tubing has to have a ¼″ air space between it and the supporting surface. All hardware has to be corrosion proof.

The EMT system must be installed so that it is free from possible physical damage. This means during the installation and after. For example, the system can't be installed in a workshop where a lot of rough mechanical work will be taking place.

However, this is a judgment that must be made by you and the inspector together. The key to approval is the word severe. EMT is rugged but it is not bullet proof or even hammer proof. When there is a possibility of severe damage, don't use EMT.

The most demanding rules, though, have to do with conduit bending, correct sizing, and calculating how many conductors will fill the pipe. These topics are covered further later in this chapter and in Chapter 6.

AREA—SQUARE INCHES

In Table 5-1 under the heading of Area-Square Inches appear the numbers you'll be using most when you perform pipe calculations.

Notice the column entitled Total 100%. This refers to the full area of the pipe's internal cross section. How is this related to column 2 (Internal Diameter Inches)?

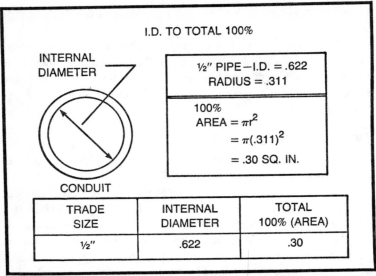

Fig. 5-5. The relationship between the internal diameter columns and total 100% columns is πr^2.

The area of a pipe in square inches is calculated from the formula πr^2. The diameter of ½″ pipe is shown as .622 square inch. Th radius r is one-half of the diameter, or .311. Multiplying $(.311)^2$ by 3.14 (π) yields .30. The answer is the value of the total 100% area in square inches (Fig. 5-5).

As you can see, the column (Total 100%) is a handy way to determine the area without going through the calculation each time.

There are eight more columns in the table. Fortunately, you'll usually be using only one of the eight: column 4 (over 2 Cond. 40%) under the heading of Not Lead Covered. (See Figs. 5-6 and 5-7.)

You should consult column 4 when you are using two or more non-lead covered wires in a pipe. If there are only two wires, then you should refer to column 3. If there is only one wire you should refer to column 5.

The figure 40% means that you cannot fill the pipe to more than 40% of its area. For example, take the ½″ pipe again. Total 100% is .30 square inches. Multiplying .30 by 40% yields .12 (square inches). Column 4 offers a handy method for getting the 40% figure without having to perform the calculations.

For EMT, you are not permitted to use a size larger than 4″. This further reduces the extent of the portion of the table you'll be interested in.

As mentioned, most of the time you'll only use the 40% figure. In addition, you'll hardly ever use more than that first little block that shows the 40% square inch area figure for the trade sizes from ½" through 1½".

Don't ignore the other columns, though, because on occasion you'll need them.

The reason you are not permitted to fill the pipe with conductors to more than 40% of its area should be obvious. It is hazardous to pack current carrying wires too tightly together. The temperature inside the pipe makes a fire a distinct possibility.

THREADING PIPE

Rigid metal conduit is threadable and can be screwed into boxes, fittings, couplings and connectors to create a rugged permanent installation. When you work with rigid metal pipe, you will use a threading machine and you will become involved in all the facets of pipe threading (Fig. 5-8).

The Code will NOT permit you to thread EMT. The walls are too thin. Since you can't thread EMT, all boxes, fittings, couplings, and connectors in the system must be installed without the benefit of threads.

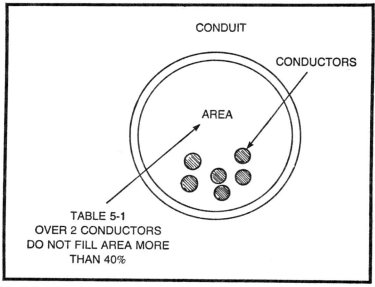

Fig. 5-6. A pipe with more than two conductors cannot be filled to more than 40% of its cross-sectional area.

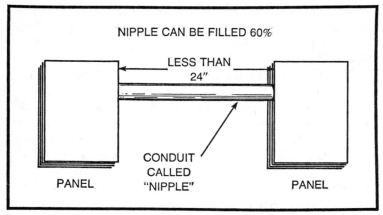

NIPPLE CAN BE FILLED 60%

LESS THAN 24"

CONDUIT CALLED "NIPPLE"

PANEL

PANEL

Fig. 5-7. The exception to the 40% of cross-sectional area regulation: a pipe shorter than 24 inches and used as a nipple between boxes can be filled to 60%. (See column 4 of Table 5-1.)

You must use threadless connectors (Fig. 5-9). There are many varieties of these connectors. Some common ones are set screw, compression, indenter, tap-on and squeeze types. Threadless connections are required to be sturdy, tight and permanent. They must be able to maintain good electrical grounds, mechanical rigidity and strength.

The EMT connections are made with a pipe wrench, not with slip-joint pliers, and must be tightened sufficiently to remain intact for years and years without attention.

EMT marries well with rigid metal conduit. Often during a job it will be advantageous to use both types of conduit in the same

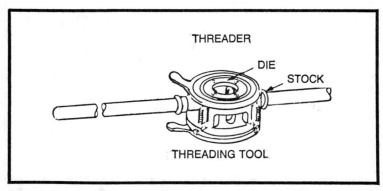

THREADER

DIE

STOCK

THREADING TOOL

Fig. 5-8. To insure permanent and safe connections, rigid pipe is threaded and installed tight.

system. In this situation you will use adapters to join the EMT components to the rigid metal components.

For example, suppose an adapter has one unthreaded end and one threaded end (supplied with a locknut). The unthreaded end is attached to the EMT in the usual fashion and the threaded end is secured (with the locknut) to a box at a knockout opening.

EMT SUPPORTS

U-channel hangers are a popular EMT support. U-channels are very handy in rooms like kitchens, basements, etc., which are fitted with lift-out ceilings. The lift-out panels offer accessible space against the ceiling for installing the EMT.

A large number of spring clips, conduit clamps and rod hangers are available to secure the EMT to the U-channel supports (Fig. 5-10).

The Code requires that EMT be fastened at 10-foot intervals along spans and at a point within 3 feet of each junction box, outlet box, cabinet or fitting.

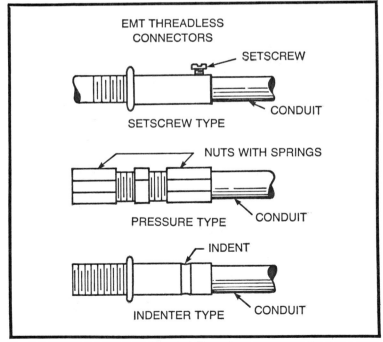

Fig. 5-9. EMT is too thin to be threaded. Threadless connectors like these must be used. They are safe for EMT installations.

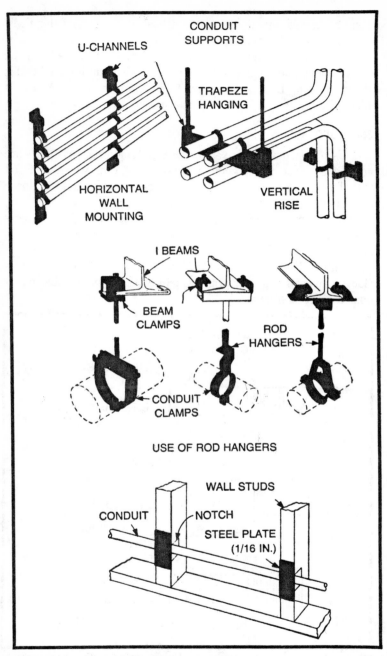

Fig. 5-10. Conduit can be supported in various ways.

When the supporting surface is wood, attachment is easy. There are a number of pipe strap types you can use with wood screws securing them into position (Fig. 5-11).

Masonry surfaces need a different type of screw. You drill holes into the masonry and install lead expansion anchors. Machine screws are turned into the lead to hold the pipe straps in place. Wooden anchors are prohibited since they may eventually dry out and loosen in the hole.

The illustrations show many different methods for supporting the EMT. Each method has its own virtues.

CONDUIT BENDING

Have you ever watched an expert electrician bend pipe? He makes it look easy. When you try to do it, it's not so easy. There is a lot more to it than there appears to be at first glance.

There are bends and offsets. There are front bends, back bends and—when you make an error—rebends.

Bends have to be exact in length and curvature. There cannot be any creases on the inside of the bend, and no injury to the tubing is permitted. Furthermore, the internal diameter of the pipe must be maintained unvaryingly around all corners. And there are plenty of corners! EMT bends easily but bending it correctly is another matter.

A good installation is a pleasure to behold. Bad bending does not look at all professional and often will not pass inspection. If an inspector sees a bad bending job, you can be sure the rest of your work is going to receive close scrutiny. That's for sure.

You must make a major effort to develop good bending techniques. The first of the important electrician's techniques we discussed was electrical connection to terminals. This is the second technique to concentrate on. Your bends should be straight and

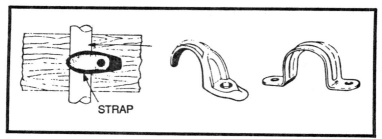

STRAP

Fig. 5-11. Conduit can readily be secured to wooden beams with straps and clamps like these.

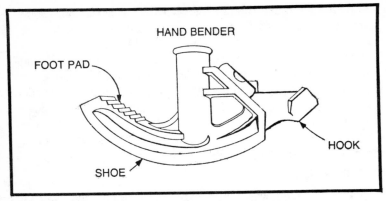

Fig. 5-12. A hand bender looks like a piece of a wheel.

square, never crooked, out of line or creased. The ability to bend conduit properly has monetary importance. The following pages will cover many of the bending techniques but more important is your dedication to the care and effort you make with the pipe.

HAND BENDING

Most of the time you'll be doing your bending by hand. For this, electricians use a gadget that looks like part of an automobile wheel. It's called a hand bender (Fig. 5-12).

It is also called a hickey (Fig. 5-13). To be exact, a hand bender as shown in the illustration is used for shaping EMT in sizes of ½", ¾", and 1". Each size gets a separate bender. A bender works properly only with the size of pipe it was designed to accommodate.

If you have a hand bender for ¾" EMT, it is also a bender for ½" rigid metal conduit. This is because the outside diameter of ¾" EMT is about equal to the outside diameter of ½" rigid metal conduit.

You can do the same thing with the benders of other sizes. The sizes of conduit which a bender can accommodate are marked conspicuously on the tool itself. So there should be no confusion.

Hand benders can be purchased that will bend EMT of trade sizes up to 2".

The hickey, to be precise, is designed to make offset bends in 1¼", 1½" and 2" EMT. In practice the words hickey and hand bender are often used interchangeably.

Hand benders are not equipped with handles when purchased. The buyer must provide his own. The most popular choice for a handle is a short length of rigid metal pipe.

Figure 5-14 shows a piece of conduit being shaped by a hand bender. The bender with a handle attached is hooked over the conduit. Pressure is applied to the handle while a foot is set upon the conduit. The leverage of the handle allows you to bend the conduit with only ordinary effort. The bend is developed in accordance with the curve of the bender.

Bends are made mostly by eye and it takes a bit of practice to get the bend made correctly. After a while as you get the feel of the bending tool and the way that various sizes and types of pipe take a bend, you'll be able to perform the operation easily.

The hickey is used by the electrician as if it were an extension of his hand and foot, but he often turns to mechanical and hydraulic benders to bend the larger sizes of pipes.

LARGER BENDERS

EMT sizes larger than 1″ are best bent with a larger hydraulic bender. You can use a hickey to make offset bends in the larger pipe, but for a good 45- or 90-degree bend, a bender as shown in Fig. 5-15 is recommended.

This type of bender operates easily and the manual pressure you exert is applied to a lever with a ratchet. You simply keep pumping the lever and the machine bends the pipe quickly and exactly. Anyone can do it, since not much technique is required. All the guesswork is removed. The pipe is placed between the pipe supports and the shoe. The clamp holds the pipe snugly against a shoe of the correct size. Instead of having a number of different

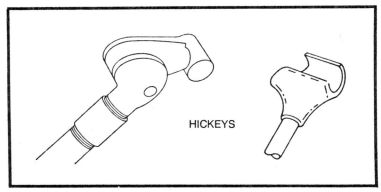

HICKEYS

Fig. 5-13. A hickey is designed to make offset bends and is different from an ordinary bender.

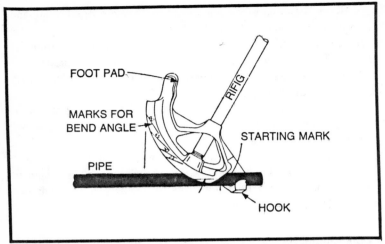

Fig. 5-14. The hook is attached to the pipe at the starting mark. The handle is pulled at the same time your foot is applying pressure to the foot pad.

benders for different sizes of pipe, you change the shoe to match the pipe.

The bender is fitted with a gauge that tells you the angle of the bend. You keep pumping the lever with short powerful strokes. The ratchet allows the lever to return after each stroke. You continue to pump until the gauge tells you that the pipe is bent to the desired angle.

POWER HYDRAULIC BENDERS

The bending of large pipe requires a great amount of pressure. The manual benders would tax your strength. A power hydraulic bender is needed. The hydraulic pressure is derived from a motor instead of from human muscles.

When bending large sizes, slide the pipe between the bending shoe and the two pipe supports. The pipe supports of the large benders come in different sizes to hold the different sizes of pipe.

The pipe supports are attached to the frame by two pipe support pins.

The RAM unit does the actual bending. It is composed of a cylinder and a ram. Its end is installed with a bending shoe and a shoe support.

The bending shoe imparts the angle. In Fig. 5-15 the angle is 90 degrees. The pumping is begun once the pipe is securely in place. Hydraulic oil is forced into the cylinder.

The enormous hydraulic pressure forces the ram out of the cylinder and presses the bending shoe against the pipe. The pipe bends and forms around the bending shoe. It assumes the curvature of the shoe.

After the bend is made, a valve is opened and the oil flows out of the cylinder and back into the pump.

THE RADIUS OF THE BEND

The radius of the bend is as important a consideration as the angle. The situation is confusing because, at first thought, one 90-degree angle would seem to be the same as another 90-degree angle.

In fact, an angle does not have a radius. It is two lines that intersect. If we consider 90 degrees, the two lines form a right angle with no radius in sight. A radius is a dimension of a circle; an angle isn't. However, in a 90-degree bend, the angle isn't an actuality. The pipe isn't formed into an intersection with two distinct sides. The radius is in the pipe and the bend lies in the nook of the 90-degree angle.

As you can see from Fig. 5-16, you can have any number of radii lying in the angle. It is all according to the size of the circle you want to nestle into the angle.

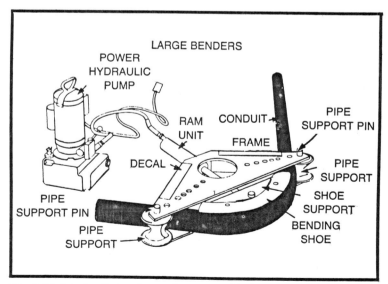

Fig. 5-15. A hydraulic bender can be operated by hand or by a motor.

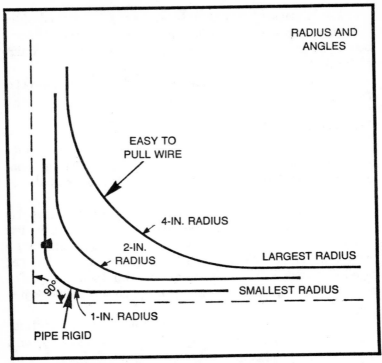

Fig. 5-16. Any number of radii can be nestled into a 90-degree angle. Don't confuse radius and angles.

The radius has a lot to do with the ease with which you'll be able to pull wire through the pipe. If the radius gets too small the wire won't pull easily. A large radius eliminates the sharp turns which make wire-pulling toilsome.

From long experience the best radius for each size of pipe has been calculated. The Code decrees that each pipe have a certain size of radius.

The Code book shows the proper radius (in inches) for bending various sizes of conduit in Table 346-10. If you bend correctly, these radii will be almost automatic. The bending equipment was designed with these radii in mind.

ONE SHOT VERSUS MULTIPLE BENDS

A one-shot bend means that you get the pipe in position for the bend carefully, then apply the pressure. The bend ends up in one continuous curve (Fig. 5-17).

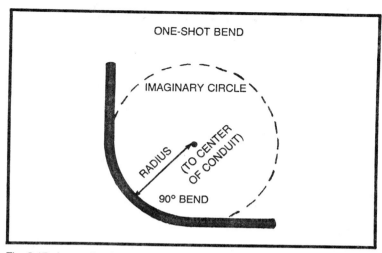

Fig. 5-17. A one-shot bend is measured to the center of the conduit.

A multiple-type bend is not one continuous curve. It consists of a number of small-angle bends which add up to the final desired angle. Instead of one continuous curve, it is composed of a number of small bends with a short length of straight pipe between each bend (Fig. 5-18).

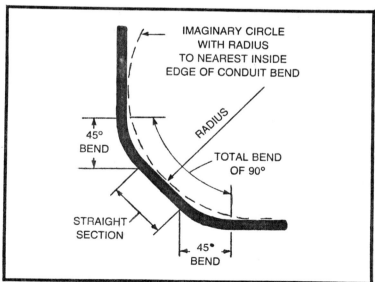

Fig. 5-18. A multiple bend is measured to the inner edge of the conduit to allow a bit more space.

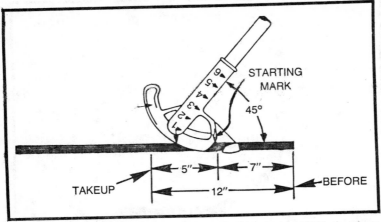

Fig. 5-19. To start the 12-inch, 90-degree bend, you measure off 7 inches from the end of the pipe and make a mark.

In the one-shot bend, the radius is measured from the center of the pipe to the center of the shoe. According to Code you cannot bend a pipe with a smaller radius than the Code's table advises.

For the multiple bend, a slightly larger radius is required. The measurement is made, not from the center of the pipe as in the one-shot, but from the inner edge.

For example, in a one-shot bend of 1" pipe, the Code's table calls for a radius of 5¾". In a multiple-type bend you'll find that the radius measures out to 6".

The multiple bend should have a slightly larger radius to help out in the wire pulling. All the little bends create a number of creases that are not present in a one-shot bend.

Another consideration in the bending of pipe is the insulation on the wire you are using. The Code's table takes into account only wires with ordinary insulation. Should you use special types of insulated wire, you might need larger bends. Check out this factor on all jobs.

BENDING TECHNIQUES

Pipe comes in standard lengths of 10 feet. A close look at the pipe shows that many manufacturers have marked the pipe at 1-inch intervals all along its length. The markings are time-saving aids for use when correct angles and radii have to be formed.

Let's examine the thinking you are to go through as you make a 90-degree bend that extends 12 inches above the running level of the pipe. The 12 inches is composed of the actual bend and the pipe that connects to a fitting.

It's called a 12-inch, 90-degree bend, but the measurement is straight (Fig. 5-19). The 12-inch, 90-degree bend uses more than 12 inches of pipe. You should always remember that the radius of the bend decides how much pipe will actually be used.

A pipe of ¾″ size is bent with a ¾″ EMT bender and has a "takeup" of 5 inches. The takeup is the important measurement. The shoe accounts for more than 5 inches but the amount of extra pipe is not important. The takeup is what is counted.

The 5 inches of takeup and the 7 inches of straight pipe add up to the required 12 inches.

The radius is automatically correct as long as it conforms to the curve of the shoe. You can make a one-shot bend with the correct angle and radius every time without a mistake (Fig. 5-20).

When you need a larger radius, you can take a number of bites and make a multiple-type bend.

To sum up, if you want a 12-inch, 90-degree bend, you measure 7 inches from the end of the pipe and make a mark. The bender is placed on the conduit with the inside of the hook at the mark. Place your foot on the other side of the bender and pull the handle slowly and surely until you have a smooth 90-degree bend.

THE BACK BEND

After you make the first bend, frequently you have to make a second bend that is a mirror image of the first. This is called a back bend.

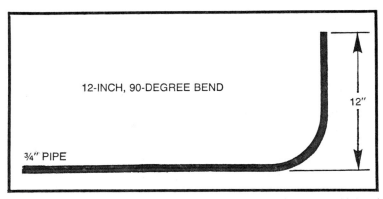

Fig. 5-20. If you apply your pressure carefully, you'll produce a smooth bend without creases.

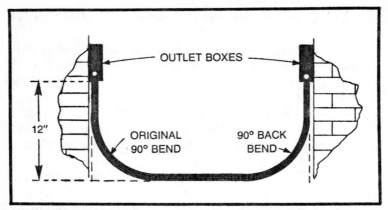

Fig. 5-21. A back bend is commonly used when two outlet boxes face each other across a room.

These twin bends often occur as you attach pipe between two outlets on facing walls. Figure 5-21 shows a typical installation.

The first bend, in this case on the left side of the room, is accomplished exactly as described in the last section (Bending Techniques). It is a 12-inch rise that consists of 7 inches of pipe sticking straight up and 5+ inches of pipe in the curve of the bend. The pipe projects into an outlet box that is situated 12 inches above the floor line. That is why the 12-inch bend is so important. It is a commonly used bend.

The first bend is made at the end of the pipe. No additional measurements are needed, but you should mark the pipe 7 inches from the end for the bending hook. That was easy, but complications can arise with the back bend. Chances are that the bend cannot be measured from the other end of the pipe. The pipe is 10 feet long and the room has a width that doesn't accord conveniently.

You compute the back bend with a plumb bob and chalk. Hang the plumb bob on the first bend and measure the distance from the bob to the beginning of the rise of the bend. This measurement is called X. Allow for the distance between the bob point and the wall. It is about ½ inch.

Next, on the opposite wall beneath the outlet box, measure off a distance the equivalent of X in the direction of the X.

Between the two X-distances is Y. Measure Y exactly. It tells you the length of the straight pipe required to join the two bends.

Now mark Y off on the pipe. One side of Y is the beginning of the first bend. At the other end of Y the back bend rise will occur.

Take the bender and place the end of the shoe opposite the hook at the point where Y ends (Fig. 5-22). Then run the shoe carefully over the pipe until the hook touches and attaches. Now make your 90-degree bend. Saw off any excess pipe that rises above the outlet box.

Should there not be enough pipe, attach another piece of pipe to the original, mark off Y, and perform the back bend.

Back bends are a little tricky, but they must become a part of your technique.

REBENDING

Sometimes when bending a pipe you'll crease or kink it so severely that it can't be used as is. But you can saw off the crushed piece and use the remainder when you need a shorter length.

At other times you can bend a bad angle or inflict defects and still use the piece for your original purpose. With a good rebend you can restore the pipe to a useful condition.

Just make sure that on a second or third rebend the walls of the pipe haven't suffered metal fatigue to the extent that they are no longer strong enough for the pipe to be used with safety.

When you make a decision to rebend, you should first determine what went wrong before. You do not want to repeat your mistake. Make sure you do it right the second time.

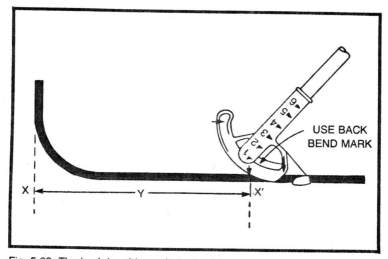

Fig. 5-22. The back bend is made by hooking the bender onto the pipe at a distance of Y.

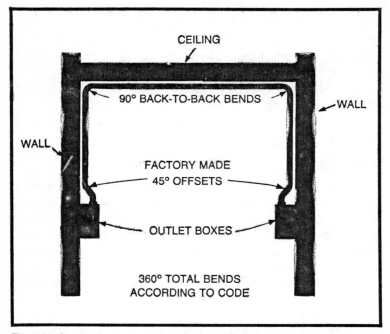

CEILING

90° BACK-TO-BACK BENDS

WALL

WALL

FACTORY MADE
45° OFFSETS

OUTLET BOXES

360° TOTAL BENDS
ACCORDING TO CODE

Fig. 5-23. Small offsets are used at outlet boxes. You can purchase pre-formed pieces of offset pipe.

The straightening of the pipe is best accomplished by sliding the rigid metal handle of the bender over the pipe. With one good strong sweep, you should be able to straighten the piece.

Then you are ready to rebend.

OFFSETS

An offset bend is really two bends. One takes the pipe out of line with the run and the second returns it parallel to the run.

Offsets can be small, as in the case of offsetting pipe from a wall to a knockout hole in an outlet box (Fig. 5-23). Offsets can be large, as they are when used to "kick over" a 4 × 8-inch joist.

The small offsets are usually done by eye. Even if you err slightly, the offsets will still appear to be straight and square.

Some electricians do not bother with small offsets and just run the pipe into a box. This is not good technique and will not win him any pats on the back.

Since offsets need two bends that conform with each other, they are more difficult to form than a single bend. Let's examine the

116

step-by-step procedure for bending an offset to kick over a 4 × 8-inch joist (Fig. 5-24).

The pipe runs directly at the side of the joist. Measure the distance between the end of the last 10-foot section of pipe and the joist. Suppose the distance is 14 inches.

Deduct 7 inches from the measurement. The 7 inches represent the takeup of the bender and one-half of the height of the 4-inch-high beam. The bender takeup is 5 inches and the half-height of the beam is 2 inches.

Each of the two bends will be bent at a 45-degree angle. One bend will carry the pipe above the joist and the second will restore it to its previous direction but at a higher elevation.

The two bends form a triangle. The three angles are of 45, 45 and 90 degrees, respectively. The distance from the beginning of the bend to the joist is 5 inches. The distance up the joist is 5 inches and the run of pipe between the two bends is 7 inches.

The bends are made by marking the pipe according to your original measurement. Attach the shoe and hook at that point and make a 45-degree bend.

Next turn the conduit around in the hook, slide up the pipe until you have an offset depth equal to the joist height of 4 inches. Make a second 45-degree bend in the other direction.

This is easy to judge with the depth scale on the bender's handle holder, or with a rule.

THE SADDLE

A saddle is really a double offset and requires four bends. The single offset is used when you have to run a pipe above a series of beams.

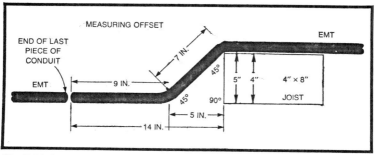

Fig. 5-24. Large offsets, such as are used for crossing over joists, are formed from two 45-degree bends.

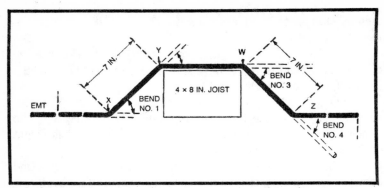

Fig. 5-25. A saddle bend is nothing more than two complete offset bends.

When only one beam is to be surmounted, it might be best to kick over the beam and then return the pipe to its original line of run.

The way it is done is to take a 45-degree kick over the beam, then take a 45-degree dip back into line. This type of offset is called a saddle (Fig. 5-25).

The step-by-step procedure for a saddle starts with a regular offset as described previously. After the offset is made, the pipe is laid across the beam to be saddled. The pipe is then marked by eye on the other side of the beam where the second bend is to be installed.

Once marked, the hook is attached and a 45-degree bend is made downward—to return the pipe to its original line of run.

The depth of the offset 4 inches is set according to the scale on the handle. The hook is attached to the spot indicated and the pipe is bent back to the original line.

An excellent saddle can be made with four 45-degree bends and twin legs for straddling the beam.

ABOUT ANGLES

It stands to reason that the sharper the angle of the pipe, the harder it will be to pull wire through. Wire may be pulled more easily through a 45-degree bend than a 90-degree bend.

The Code is specific about bending angles. It specifies that the bends of each run cannot have a combined total in excess of 360 degrees in bends (Fig. 5-23). A run is the span between two outlets, between two fittings or between an outlet and a fitting.

The specified total could consist of four 90-degree bends, two 90-degree bends and four 45-degree bends, or whatever. The more bending, the more wire pulling problems and the less safe the installation is.

During pulling operations, insulation can get stripped off and wires can cut into one another. Electric currents can get mixed. These faults can be dangerous.

A pipe can develop burrs, creases and other rough spots as it is being cut. Insulation can get hung up on these spots and jeopardize the safety of the installation.

Great care must be taken to ream all conduit ends and to smooth out any spots where the insulation might possibly hang up. It is good technique to inspect the pipe carefully and file off any semblance of a rough edge.

When you start pulling the wire, you'll be happy you smoothed the rough places, and you'll be quite unhappy if you missed a burr or two.

CONDUCTORS

Electrical conductors can be made of copper, aluminum or copper-clad aluminum. They are all metals with an abundance of free electrons which will move easily when electric pressure is applied.

GENERAL WIRING

Copper is the commonest type of conductor and most of the discussions will be about copper. Just remember that the other two types are also used widely, and their characteristics, while not quite equaling those of copper, are satisfactory.

When conductors are called wire, it usually means they are insulated. However, the term wire is also used in reference to uninsulated conductors.

Insulation comes in many different types. When you choose a wire, you choose with the insulation in mind. Some of your thoughts are: will the location be wet or dry; will the environment be hot or cold; will chemicals or oil be nearby; will the installation be under ground or above ground. Insulations are available for all of these applications.

When you think about wire, there are seven main characteristics to be considered.

1—Size
2—Kind of metal
3—Resistance

4—Ampacity and voltage drop
5—Weight
6—Type of insulation
7—Tensile strength

WIRE SIZES

Wire is sized in two ways. One way is by circular mils. A mil is a thousandth of an inch. A circular mil is related to a square mil (Fig. 6-1).

The second way of sizing wire is by the American Wire Gage (called AWG) standard. AWG sizes extend from 18 to 4/0 in the NEC tables (Table 6-1).

The sizes of wires—as we have discussed them—are shown in columns 1 and 2 of the Code book's Table 8, Properties of Conductors. The sizes refer to the wires' cross-sectional areas.

Column 1 lists the AWG number along with the size in MCM units. MCM means 1000 circular mils.

As you follow down the two columns, the wires get larger (Fig. 6-2). Number 18 is the smallest size rated by the AWG numbers and 4/0 (0000) is the largest.

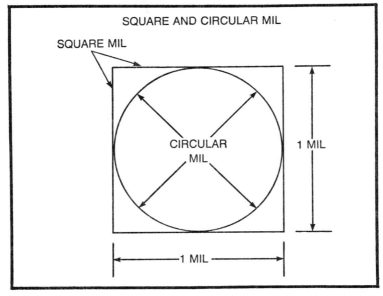

Fig. 6-1. A square mil is larger than a circular mil. A square mil is 1/1000 of an inch while a circular mil is about 22% smaller.

Table 6-1. Properties of Conductors. (National Electrical Code)

Size AWG, MCM	Area Cir. Mils	Concentric Lay Stranded Conductors		Bare Conductors		DC Resistance Ohms/M Ft. At 25°C, 77°F.		
		No. Wires	Diam. Each Wire Inches	Diam. Inches	*Area Sq. Inches	Copper Bare Cond.	Copper Tin'd. Cond.	Aluminum
18	1620	Solid	.0403	.0403	.0013	6.51	6.79	10.7
16	2580	Solid	.0508.	.0508	.0020	4.10	4.26	6.72
14	4110	Solid	.0641	.0641	.0032	2.57	2.68	4.22
12	6430	Solid	.0808	.0808	.0051	1.62	1.68	2.66
10	10380	Solid	.1019	.1019	.0081	1.018	1.06	1.67
8	16510	Solid	.1285	.1285	.0130	.6404	.659	1.05
6	26240	7	.0612	.184	.027	.410	.427	.674
4	41740	7	.0772	.232	.042	.259	.269	.424
3	52620	7	.0867	.260	.053	.205	.213	.336
2	66360	7	.0974	.292	.067	.162	.169	.266
1	83690	19	.0664	.332	.087	.129	.134	.211
0	105600	19	.0745	.372	.109	.102	.106	.168
00	133100	19	.0837	.418	.137	.0811	.0843	.133
000	167800	19	.0940	.470	.173	.0642	.0668	.105
0000	211600	19	.1055	.528	.219	.0509	.0525	.0836
250	250000	37	.0822	.575	.260	0.431	.0449	.0708
300	300000	37	.0900	.630	.312	.0360	.0374	.0590
350	350000	37	.0973	.681	.364	.0308	.0320	.0505
400	400000	37	.1040	.728	.416	.0270	.0278	.0442
500	500000	37	.1162	.813	.519	.0216	.0222	.0354
600	600000	61	.0992	.893	.626	0.180	.0187	.0295
700	700000	61	.1071	.964	.730	.0154	.0159	.0253
750	750000	61	.1109	.998	.782	.0144	.0148	.0236
800	800000	61	.1145	1.030	.833	.0135	.0139	.0221
900	900000	61	.1215	1.090	.933	.0120	.0123	.0197
1000	1000000	61	.1280	1.150	1.039	.0108	.0111	.0177
1250	1250000	91	.1172	1.289	1.305	.00863	.00888	.0142
1500	1500000	91	.1284	1.410	1.561	.00719	.00740	.0118
1750	1750000	127	.1174	1.526	1.829	.00616	.00634	.0101
2000	2000000	127	.1255	1.630	2.087	.00539	.00555	.00885

*Area given is that of a circle having a diameter equal to the over-all diameter of a stranded conductor.

The values given in the table are those given in Handbook 100 of the National Bureau of Standards except that those shown in 8th column are those given in Specification B33 of the American Society of Testing and Materials, and those shown in the 9th column are those given in Standard No. S-19-81 of the Insulated Power Cable Engineers Association and Standard No. WC3-1969 of the National Electrical Manufacturers Association.

The resistance values given in the last three columns are applicable only to direct current. When conductors larger than No. 4/0 are used with alternating current, the multiplying factors in Table 9 compensate for skin effect.

The MCM measurement takes over at 250. The wire continues to increase in size. The last entry in the column is 2000 MCM.

Column 2 shows the circular mil area of #18 to be 1620. The area of the 2000 MCM wire is 2,000,000 circular mils. Note that starting with wire #250 the circular mil result is exactly 1000 times the wire number.

You can measure the size of a wire with a micrometer (Fig. 6-3). The micrometer measures the diameter of the wire in mils. You then convert the diameter to a radius, square the quantity and multiply by pi.

An easier way to measure wire is to use a standard wire gage, a disk of metal with precisely bored holes. The gage in Fig. 6-4 accommodates sizes 33 to 0 (single zero).

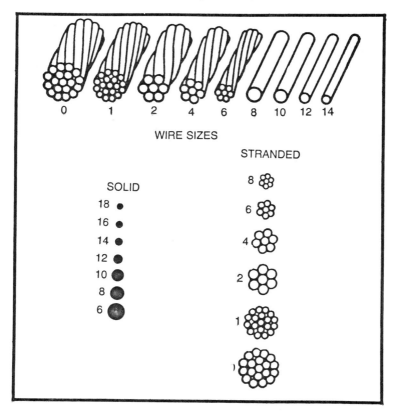

Fig. 6-2. Wires are solid from #18 down to about #6. Larger sizes use standard wire to aid handling characteristics. Sizes 2, 4 and 6 use 7 strands while larger sizes use 19 strands.

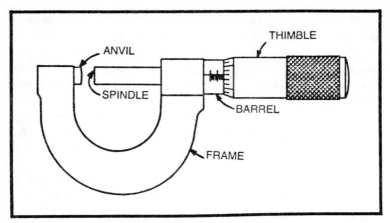

Fig. 6-3. A micrometer is one of the measuring devices. It gives you the diameter size of a wire.

CURRENT IN WIRE

The wire size is obviously related to the number of free electrons in the conductor. A fat wire has an abundance of free electrons packed away in its obesity. A thin wire, because of its leanness, has far fewer electrons available for movement.

The more free electrons, the heavier the current carrying capability of the wire. It is apparent that thick wire can carry more current than thin wire.

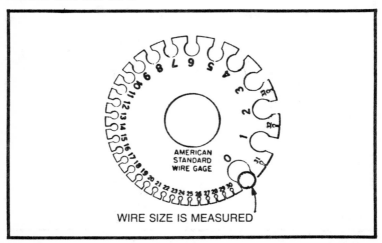

Fig. 6-4. An easier way to measure wire size is with a wire gage with punched holes.

If you try to force a lot of current into a thin wire, the wire will heat up. As a result the Code stipulates how much current (in amperes) should be introduced into each size of wire.

If you look at Table 6-2, the Allowable Ampacities of Insulated Copper Conductors is shown. Column 1 is the wire size in AWG and MCM. It progresses from #18 wire to #2000 wire.

Alongside each wire size is the amount of current you permit to be entered into a wire of this thickness.

There are columns for many kinds of wires. The copper is the same in all of the wires, but the insulation is different. The highest temperature that each type of insulation can endure is shown at the top of the columns.

Typical wires that an electrician pulls are types like TW and THW in the 60° and 75°C columns. You can see that wire sizes #14, #12, #10 and #8 each has its own ampacity. Memorize those numbers. Fifteen, 20, 30, and 40 amps, respectively. You can consider #6 wire as being suitable for 50 amps.

These are the wires you'll be pulling most on residential and small commercial jobs.

These currents are allowed for all of the common voltages, such as 120, 240 and 208 volts. Don't get confused between the voltage applied and the ampacity. The ampacities in the table apply the same to all of the voltages.

WIRE RESISTANCE

In a thin wire that has a limited amount of free electrons, the ampacity is limited. More voltage is required to push electrons through the wire. It is said, therefore, that a thin wire has a high resistance to electron flow.

A thick wire has a large quantity of free electrons; less voltage is required to push electrons through the wire. Therefore, a thick wire has a low resistance to electron flow.

These principles can be observed in Ohm's Law. If you have a resistance of 10 ohms in a 120-volt circuit, then

$$I = E/R = 120/10 = 12 \text{ amps}$$

This represents thin wire.

If you have a thicker wire that has half of the previous resistance (or 5 ohms), then

$$I = E/R = 120/5 = 24 \text{ amps}$$

Table 6-2. Permitted Ampacities of Various Conductors (National Electrical Code

Allowable Ampacities of Insulated Conductors
Rated 0-2000 Volts, 60° to 90°C

Not More Than Three Conductors in Raceway or Cable or Earth
(Directly Buried), Based on Ambient Temperature of 30°C (86°F)

Size	Temperature Rating of Conductor. See Table 310-13								AWG
	60°C (140°F)	75°C (167°F)	85°C (185°F)	90°C (194°F)	60°C (140°F)	75°C (167°F)	85° (185°F)	90°C (194°F)	
AWG MCM	TYPES RUW, T, TW, UF	TYPES FEPW RH, RHW RUH, THW, THWN, XHHW, USE, ZW	TYPES V, MI	TYPES TA, TBS, SA, AVB, SIS, FEP, FEPB, RHH, THHN, XHHW°	TYPES RUW, T, TW, UF	TYPES RH, RHW, RUH, THW, THWN, XHHW, USE	TYPES V, MI	TYPES TA, TBS, SA, AVB, SIS, RHH, THHN XHHW°	AWG MCM
	COPPER				ALUMINUM OR COPPER-CLAD ALUMINUM				
18	....	....	....	21	....	....	....	....	....
16	....	....	22	22	....	....	....	....	....
14	15	15	25	25	....	....	....	....	....
12	20	20	30	30	15	15	25	25	12
10	30	30	40	40	25	25	30	30	10
8	40	45	50	50	30	40	40	40	8
6	55	65	70	70	40	50	55	55	6
4	70	85	90	90	55	65	70	70	4
3	80	100	105	105	65	75	80	80	3
2	95	115	120	120	75	90	95	95	2
1	110	130	140	140	85	100	110	110	1
0	125	150	155	155	100	120	125	125	0
00	145	175	185	185	115	135	145	145	00
000	165	200	210	210	130	155	165	165	000
0000	195	230	235	235	155	180	185	185	0000
250	215	255	270	270	170	205	215	215	250
300	240	285	300	300	190	230	240	240	300
350	260	310	325	325	210	250	260	260	350
400	280	335	360	360	225	270	290	290	400
500	320	380	405	405	260	310	300	330	500
600	355	420	455	455	285	340	370	370	600
700	385	460	490	490	310	375	395	395	700
750	400	475	500	500	320	385	405	405	750
800	410	490	515	515	330	395	415	415	800
900	435	520	555	555	355	425	455	455	900
1000	455	545	585	585	375	445	480	480	1000
1250	495	590	645	645	405	485	530	530	1250
1500	520	625	700	700	435	520	580	580	1500
1750	545	650	735	735	455	545	615	615	1750
2000	560	665	775	775	470	560	650	650	2000

Table 6-2. Con't.

Allowable Ampacities for Insualted Conductors
Rated 110 to 250°C

Not More Than Three Conductors in Raceway or Cable
Based on Ambient Temperature of 30°C (86°F).

Size	Temperature Rating of Conductor. See Table 310-13								Size
	110°C (230°F	125°C (275°F)	150°C (302°F)	200°C (392°F	250°C (482°F)	110°C (230°F)	125°C (257°F)	200°C (392°F)	
AWG MCM	TYPES AVA, AVL	TYPES AI, AIA	TYPE Z	TYPES A, AA, FEP, FEPB, PFA	TYPES PFAH, TFE	TYPES AVA, AVL	TYPES AI, AIA	TYPES A, AA	AWG MCM
	COPPER				NICKEL OR NICKEL-COATED COPPER	ALUMINUM OR COPPER-CLAD ALUMINUM			
14	30	30	30	30	40	....	....	....	....
12	35	40	40	40	55	25	30	30	12
10	45	50	50	55	75	35	40	45	10
8	60	65	65	70	95	45	50	55	8
6	q80	q85	90	95	120	60	65	75	6
4	105	115	115	120	145	80	90	95	4
3	120	130	135	145	170	95	100	115	3
2	135	145	150	165	195	105	115	130	2
1	160	170	180	190	220	125	135	150	1
0	190	200	210	225	250	150	160	180	0
00	215	230	240	250	280	170	180	200	00
000	245	265	275	285	315	195	210	225	000
0000	375	310	325	340	370	215	245	270	0000
250	315	335	....	....	....	250	270	....	250
300	345	380	....	....	....	275	305	....	300
350	390	420	....	....	....	310	335	....	350
400	420	450	....	....	....	335	360	....	400
500	270	500	....	....	....	380	405	....	500
600	525	545	....	....	....	425	440	....	600
700	560	600	....	....	....	455	485	....	700
750	580	620	....	....	....	470	500	....	750
800	600	640	....	....	....	485	520	....	800
1000	680	730	....	....	....	560	600	....	1000
1500	785	....	....	....	....	650	....	....	1500
2000	840	....	....	....	....	705	....	....	2000

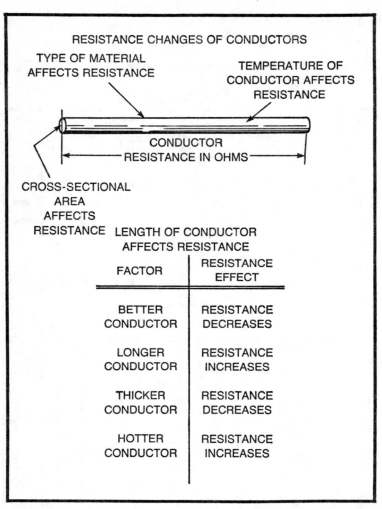

Fig. 6-5. Material, length, temperature and cross-sectional area all affect the resistance of a conductor.

You can pass twice as much current with the same voltage.

Suppose you have a piece of wire with a resistance of 10 ohms, but you put it in a 240-volt circuit.

$$I = E/R = 240/10 \text{ amps}$$

As you can see, when you double the voltage, you double the current. Also, if you halve the wire resistance, you double the current. Remember this little exercise, you'll be using the concept continually as you wire electrical systems (Fig. 6-5).

To be exact on wire resistances, turn to Table 6-1 again. The right hand part of the table has a section called DC Resistance, Ohms/M Ft., at 25°C, 77°F.

The Ohms/M Ft. means ohms per 1000 feet. Notice that as the wire sizes in Column 1 grow thicker, the ohms become less and less. Also notice that aluminum wire has a higher resistance than copper to current flow, all the way down the chart. The aluminum, though, also has increasingly less resistance as the wire gets thicker.

The ohms are noticeably low. One thousand feet of bare #18 copper wire has a resistance of only 6.51 ohms. The resistance of 100 feet, then, is 0.651 ohms, just a bit more than half an ohm.

During World War II a copper shortage developed and silver conductors had to be used in some of the war materials.

Silver is about 5 percent more effective than copper as a conductor.

Aluminum has become a fairly common wiring material. Even though it is a slightly less efficient conductor than copper, its lower cost makes it a practical substitute. In addition, aluminum has the advantage of being much lighter in weight than copper. When aerial wires have to be strung long distances, aluminum is much easier to support.

Aluminum, as a rule, must be one size thicker than copper in order to conduct the same amount of electricity.

For example, in Table 6-2, wire type TW copper needs #14 wire to carry 15 amps. Its aluminum counterpart, on the other hand, requires #12 wire if it is to carry 15 amps.

Conductors of steel, bronze and combinations of metals are sometimes used to advantage in special installations. After all is said and done, however, copper wire is still the normal conductor of the electrical industry.

Conductors can be made up as a single solid wire, or as a number of individual bare wires twisted together. Stranding wire takes the clumsiness out of a thick wire, giving it a lot of flexibility.

Conductors in thin sizes usually consist of a single wire; large and intermediate size wires are often stranded. When you choose a wire, the handling characteristics can be important and you might want to choose a wire according to its flexibility. Conduction of solid and stranded wires is identical.

Copper wires come in three grades of hardness. They are *hard-drawn, medium hard-drawn, and soft-drawn.* Soft drawn is known as *annealed.* Most residential wire is soft-drawn. The other kinds are commercial and industrial types.

INSULATION CODING

Initials are printed on the insulation of wires. The initials stand for particular qualities of the insulation. You've already seen in Table 6-2 certain initials under certain temperatures. This means that TW is a 60°C wire and THW is a 75°C wire. The other wire initials also designate the maximum temperatures that the insulation can withstand.

What do the initials stand for? You should memorize these designations. The following are in common use.

T—thermoplastic
W—wet or dry locations
H—higher operating temperature than T
HH—higher operating temperature than H
N—nylon outer jacket
X—cross linked synthetic polymer
R—rubber
U—90° unmilled granular rubber
M—machine tool use
V—varnished cambric
A—asbestos or glass braid
L—lead sheath
UF or USE—underground use.

Take a look at the various types in Table 6-2. Note TW at 60°C. It refers to thermoplastic wire that can be used in wet or dry places. Now look at THW at 75°C. It is like TW but the H means it can withstand more heat than TW.

Now look at THHN under 90°C. It is also a thermoplastic (T) with a nylon outer jacket (N). The HH means it can withstand higher temperatures than just plain H. You can identify all of the various kinds of wire by the initials.

WIRE CROSS-SECTIONAL AREAS

It is now time to take a close look at another important table from the Code book (Table 6-3). The two-part table presents the dimensions of the various rigid conductors (wires) you might use.

Table 6-3. Dimensions of Rubber-Covered and Thermoplastic-Covered Conductors. (National Electrical Code)

Size AWG MCM	Types RFH-2, RH, RHH,*** RHW,***SF-2		Types TF, T, THW,† TW, RUH,** RUW**		Types TFN, THHN, THWN		Types**** FEP, FEPB, FEPW, TFE, PF, PFA, PFAH, PFG, PTF, Z, ZF, ZFF		Type XHHW, ZW†		Types KF-1, KF-2, AFF-2, KFF-2	
	Approx. Diam. Inches	Approx. Area Sq. In.	Approx. Diam. Inches	Approx. Area Sq. In.	Approx. Diam. Inches	Approx. Area Sq. In.	Approx. Diam. Inches	Approx. Area Sq. Inches	Approx. Diam. Inches	Approx. Area Sq. Inches	Approx. Diam. Sq. In.	Approx. Area Sq. In.
Col. 1	Col. 2	Col. 3	Col. 4	Col. 5	Col. 6	Col. 7	Col. 8	Col. 9	Col. 10	Col. 11	Col. 12	Col. 13
18	.146	.0167	.106	.0088	.089	.0064	.081	.0052	...	....	.065	.0033
16	.158	.0196	.118	.0109	.100	.0079	.092	.0066	...	....	.070	.0038
14	30 mills .171	.0230	.131	.0135	.105	.0087	.105 .105	.0087 .0087			.083	.0054
14	45 mils .204*	.0327*	.162†	.0206†					.129	.0131		
14	......											
12	30 mills .188	0.278	.148	.0172	.122	.0117	.121 .121	.0115 .0115			.102	.0082
12	45 mils .221*	.0384*	.179†	.0251†					.146	.0167		
12	......											
10	242	.0460	.168	.0224	.153	.0184	.142 .142	.159 .0159			.124	.0121
10	......		.199†	.0311†					.166	.0216		
8	328	0.0854	.245	.0471	.218	.0373	.206 .186	.0333 .0272				
8	....		.276†	.0598†					.241	.0456		
6	.397	.1238	.323	.0819	.257	.0519	.244 .302	.467 .0716	.282	.0625		
4	.452	.1605	.372	.1087	.328	.0845	.292 .350	.0669 .0962	.328	.0845		
3	.481	.1817	.401	.1263	.356	.0995	.320 3.78	.0803 .1122	.356	.0995		
2	.513	.2067	.433	.1473	.388	.1182	.352 .410	.0973 .1316	.388	.1182		
1	.588	.2715	.508	.2027	.450	.1590	.420	.1385	.450	.1590		
0	6.29,	.3107	.549	.2367	.491	.1893	.462 ...	.1676 ...	.491	.1893		
00	.675	.3578	.595	.2781	.537	.2265	.498 ...	.1974 ...	.537	.2265		
000	.727	.4151	.648	.3288	.588	.2715	.560 ...	.2463 ...	.588	.2715		
0000	.785	.4840	.705	.3904	.646	.3278	.618 ...	2.999 ...	.646	.3278		

Part one lists wire sizes from #18 to #4/0; part two lists wire sizes from #250 to #2000.

Located in the Code book next to the table of conductor dimensions is the table (Table 5-1) devoted to the demensions and cross-sectional area of conduit and tubing. The Code book places the tables close together because they are often used in concert. They both are needed when you want to fit wires into a pipe. The tables tell you which wires and how many can be inserted in accordance with Code standards.

Under each type of wire in Table 6-3 are two columns. One column lists the diameter of the wire in inches and the other lists the cross-sectional area in square inches. Most of the time you'll need the area column.

The diameter column is related to the area column by πr^2 = area in square inches. For example #6 RH wire is 0.397 inches in diameter. The radius is one-half of the diameter, or 0.1985. Thus, pi multiplied by the radius squared equals 0.1238. That is the area of #6 RH conductor in square inches.

At any rate, you'll be using the area column most of the time. For instance, you might have to put two #6 TW's and one #8 TW into a pipe.

$$\text{\#6 TW} = .0819 \times 2 = .1638$$
$$\text{\#8 TW} = .0471 \times 1 = .0471$$
$$\text{Total wire area} = .2109 \text{ sq. in.}$$

Now that you know the total area that the wires will occupy, you can determine the size of the pipe needed to satisfy code standards.

Now refer to Table 5-1. Consult the column entitled Not Lead Covered, Over 2 Cond. 40%. A ¾-inch pipe at .2100 is not quite large enough according to Code. You'll have to go to a 1-inch pipe that allows .3400. The extra capacity might come in handy in the future.

Be careful in the use of the table. There are all kinds of exceptions, markings and notations. For example THW has a dagger (†) after it. This means that any wire size followed by a dagger refers to THW. If you look, there are two listings for #8 wire; one with a dagger and the other without. The one with a dagger .0598 is THW. The other #8 wire .0471 is a non-THW type.

Size #6 wire has only one listing and no dagger. It refers to THW as well as all the other wire types in that column.

Look at the very end of the table. There are a number of notations. An important one is ***. Dimensions of RHH and RHW without outer covering are the same as THW.

When you use RHH and RHW without outer covering, just think of them as having the same dimensions as THW. This topic can crop up as part of a trick question in certificate examinations.

FLEXIBLE CORDS

Flexible cords are conductors, but they are not the kind that is installed in pipes and tubing.

You cannot use flexible cords for fixed wiring which runs through holes in walls, ceilings and floors, is attached to building surfaces, or passes within enclosed areas.

You find flexible cords with a plug on the end attached to appliances. That is the common flexible cord you are most familiar with. The cord can be thin like the one on a table radio, or thick like the one on a waffle iron.

Flexible cords are treated in Table 6-4 (reproduced from Code book Tables 400-4 and 400-5). The cords commonly range from #27 wire parallel tinsel cord down to a #12 which is used on some refrigerators and room air conditioners. You can find some range cords as large as #4. Industrial flexible cords are manufactured as large as #2.

Table 6-4(b) gives the ampacity of flexible cords. The #27 wire is allowed to carry 0.5 amps. The highest ampacity allowed in a flexible cord is the 95 amps by #2 wire.

Flexible cords come in 2- and 3-conductor varieties. The latter type is used when a ground wire is needed.

Typical lamp cords connect lamps and fixtures to outlets. They are composed of fine strands of copper and covered with a rubber insulation. Some lamp cords are further covered with cotton or rayon wrappings to enhance the insulating qualities.

Heater cords are designed to carry heavier amounts of current yet still be flexible. They are made of stranded copper with an asbestos wrap enveloping the rubber insulation. A braided cotton fabric sheathes the asbestos.

Service cords sometimes have to pass without overheating a large amount of current over long distances when they are connected to power tools, portable motors and the like. These are large-size wires covered with rubber insulation. The rubber may be furthered covered with fiberglass, paper and neoprene.

Table 6-4(a). Flexible Cords and Cables. (National Electrical Code)

Trade Name	Type Letter	Size AWG	No. of Conductors	Insulation	Nominal *Insulation Thickness AWG	Mils	Brad on Each Conductor	Outer Covering	Use		
Parallel Tinsel Cord,	TP See Note 3	27	2	Rubber	27	30	None	Rubber	Attached to an Appliance	Damp Places	Not Hard Usage
	TPT See Note 3	27	2	Thermoplastic			None	Thermoplastic	Attached to an Appliance	Damp Places	Not Hard Usage
Backeted Tensel Cord	TS See Note 3	27	2	Rubber	27	15	None	Rubber	Attached to an Appliance	Damp Places	Not Hard Usage
	TST See Note 3	27	2	Thermoplastic			None	Thermoplastic	Attached to an Appliance	Damp Places	Not Hard Usage
Bestos-Covered Heat-Resistant Cord	AFC	18-10	2 or 3	Impregnated Asbestos	18-14	30	Cotton or Rayon	None	Pendant	Dry Places	Not Hard Usage
	AFPD		2		12-10	45	None	Cotton, Rayon or Saturated Asbestos			
			2 or 3								

See Notes 1 through 9 preceding and following table.
* See Note 9.

WEIGHT OF WIRE

When wire is run inside pipe, its weight needs no consideration. If the pipe is installed and supported according to Code, and filled to the Code percentage, the wire is not going to be too heavy.

Table 6-4(b). Ampacity of Flexible Cords and Cables. (National Electrical Code)

Size AWG	Rubber Types TP, TS Thermoplastic Types TPT, TST	Rubber Types C, PD, E, EO, EN, S, SO, SRD, SJ, SJO, SV, SVO, SP Thermoplastic Types ET, ETT, ETLB, ETP, ST, STO, SRDT, SJT, SJTO, SVT, SVTO, SPT		Types AFS, AFSJ, HPD, HDJ, HSJO, HS, HSO, HPN	Cotton Types CFPD* Asbestos Types AFC* AFPD*
		A	B		
27**	0.5	..	..	..	..
18	..	7	10	10	6
17	..		12	..	..
16	..	10	13	15	8
15	..	..	..	17	..
14	..	15	18	20	17
12	..	20	25	30	23
10	..	25	30	35	28
8	..	35	40	..	..
6	..	45	55	..	..
4	..	60	70	..	..
2	..	80	95	..	..

*These types are used almost exclusively in fixtures where they are exposed to high temperatures and ampere ratings are assigned accordingly.
**Tinsel cord.
The ampacities under sub-heading A apply to 3-conductor cords and other multiconductor cords connected to utilization equipment so that only 3 conductors are current carrying. The ampacities under sub-heading B apply to 2-conductor cords and other multiconductor cords connected to utilization equipment so that only 2 conductors are current carrying.
NOTE: Ultimate Insulation Temperature. In no case shall conductors be associated together in such a way with respect to the kind of circuit, the wiring method used, or the number of conductors that the limiting temperature of the conductors will be exceeded.

On the other hand, if you are running an overhead cable to a building or pole on your farm, you'd better give the weight some consideration (Table 6-5).

According to the table, wire has a certain weight per 1000 feet that has been calculated. Size 4/0 wire weighs 640.5 pounds per thousand feet while #14 wire weighs only 12.43 pounds per thousand feet. This refers to bare copper wire; the insulation adds more weight.

If you are in doubt, you can take a length of wire—say 100 feet long—and actually weigh a coil of it. If you need to know what the weight of your run is going to be, simple mathematics will tell you.

For example, if it is a 1000-foot coil of #8 wire which weighs 55 pounds, 100 feet of it is going to weigh 5.5 pounds. Should the run be 100 feet long and you have two wires, one for the hot line and the other for the neutral, then the combined length will be 200 feet. That would be about 11 pounds of wire all told.

When you run the wire, you have to be sure that the supports will hold the 11 pounds under all conditions short of hurricane or earthquake. This means that the weight of ice and snow and the force of high wind should be considered. Perhaps you'll need a pole at the 50-foot mark to bolster the suspension.

TENSILE STRENGTH OF WIRE

Tensile strength considerations are as important as those of weight. As you increase tensile strength, you also increase cost and weight of wire. To attain greater tensile strength you must make the wire thicker.

The Code specifies the minimum size that a wire must have if it is to be suspended over appreciable distances. Weight and tensile strength are taken into account.

You are permitted to use nothing smaller than #10 wire for spans up to 50 feet. If the run exceeds 50 feet, then you must use wire no smaller than #8.

PULLING WIRE

Pulling wire is the most demanding job an electrician performs. The task probably consumes more time than any other routine of a wireman. You can reduce the wire pulling time by the use of tried-and-true techniques. Even with good techniques you'll have problems, but it'll be best if you learn early to avoid the problems incurred through carelessness.

Table 6-5. Solid Bare Copper Conductors. (Anaconda Wire and Cable Co.)

Size AWG	Wire diam., in.	Cross sectional area Cir mils	Cross sectional area Sq. in.	Weight Per 1,000 ft, lb	Weight Per mile, lb	Hard-drawn wire Min breaking strength, lb[a]	Hard-drawn wire Max resistance per 1,000 ft at 20°C, ohms[a]	Medium hard-drawn wire Min breaking strength, lb[a]	Medium hard-drawn wire Max resistance per 1,000 ft at 20°C, ohms[a]	Soft or annealed wire Max breaking strength, lb[a]	Soft or annealed wire Max resistance per 1,000 ft at 20°C, ohms[b]
4/0	0.4600	211,600	0.1662	640.5	3,382	8,143	0.05045	6,980	0.05019	5,983	0.04993
3/0	0.4096	167,800	0.1318	507.9	2,082	6,722	0.06361	5,667	0.06329	4,745	0.06296
2/0	0.3648	133,100	0.1045	402.8	2,127	5,519	0.08021	4,599	0.07980	3,763	0.07939
1/0	0.3249	105,500	0.08289	319.5	1,687	4,517	0.1011	3,730	0.1006	2,984	0.1001
1	0.2893	83,690	0.06573	253.5	1,338	3,688	0.1287	3,024	0.1282	2,432	0.1262
2	0.2576	66,370	0.05213	200.9	1,061	3,003	0.1625	2,450	0.1617	1,929	0.1592
3	0.2294	52,630	0.04134	159.3	841.2	2,439	0.2049	1,984	0.2038	1,530	0.2007
4	0.2043	41,740	0.03278	126.4	667.1	1,970	0.2584	1,584	0.2570	1,213	0.2531
5	0.1819	33,100	0.02600	100.2	529.1	1,591	0.3258	1,264	0.3241	961.9	0.3192
6	0.1620	26,250	0.02062	79.46	419.6	1,280	0.4108	1,010	0.4087	762.9	0.4025
7	0.1443	20,820	0.01635	63.02	332.7	1,030	0.5181	806.6	0.5154	605.0	0.5075
8	0.1285	16,510	0.01297	49.97	263.9	826.0	0.6533	643.9	0.6499	479.8	0.6400
9	0.1144	13,090	0.01028	39.63	209.3	661.2	0.8238	514.2	0.8195	380.5	0.8070
10	0.1019	10,380	0.008155	31.43	165.9	529.2	1.039	401.4	1.033	314.0	1.018
11	0.09074	8,234	0.006467	24.92	131.6	422.9	1.310	327.6	1.303	249.0	1.283
12	0.08081	6,530	0.005129	19.77	104.4	337.0	1.652	261.6	1.643	197.5	1.618
13	0.07196	5,178	0.004067	15.68	82.77	268.0	2.083	208.8	2.072	156.6	2.040
14	0.06408	4,107	0.003225	12.43	65.64	213.5	2.626	166.6	2.613	124.2	2.573
15	0.05707	3,257	0.002558	9.858	52.05	169.8	3.312	133.0	3.295	98.48	3.244
16	0.05082	2,583	0.002028	7.818	41.28	135.1	4.176	106.2	4.154	78.10	4.091
17	0.04526	2,048	0.001609	6.200	32.74	107.5	5.266	84.71	5.230	61.93	5.158
18	0.04030	1,624	0.001276	4.917	25.96	85.47	6.640	67.61	6.606	49.12	6.505
19	0.03589	1,288	0.001012	3.899	20.59	67.99	8.373	53.95	8.330	38.95	8.202
20	0.03196	1,022	0.0008023	3.092	16.33	54.08	10.56	43.05	10.50	30.89	10.34
21	0.02846	810.1	0.0006363	2.452	12.95	43.07	13.31	34.36	13.24	24.50	13.04
22	0.02535	642.5	0.0005046	1.945	10.27	34.26	16.79	27.41	16.70	19.43	16.45
23	0.02257	509.5	0.0004001	1.542	8.143	27.25	21.17	21.87	21.06	15.14	20.74
24	0.02010	404.0	0.0003173	1.233	6.458	21.67	26.69	17.45	26.56	12.69	26.15
25	0.01790	320.4	0.0002517	0.9699	5.121	17.26	33.66	13.92	33.49	10.07	32.97

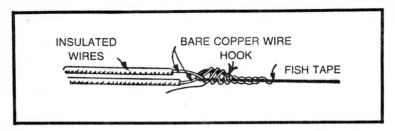

Fig. 6-6. A fish tape with a hook on the end is needed to pull wires through conduit or walls.

You just can't feed wire into a pipe or a hole in the wall and expect it to travel easily to the other end. Wire, even of the larger sizes, has a certain amount of flexibility. If you feed it into a hole, it will soon begin to diverge from the intended route. Before long it lodges against an obstruction. It will go no further. Obviously, this procedure won't work. I'm sure you know the wire has to be "fished."

Every reader at one time or another has had occasion to take a metal clothes hanger, undo it, and use it to fish wire through a wall. Even if you haven't, you know what I mean.

Each electrician has a fish tape, which is little more than a super coat hanger. A professional's fish tape is wound on a reel. It is also called "fishing wire." It is made of tempered steel wire with a rectangular cross section. It is sometimes referred to as a "snake."

The most commonly used fish wire has a width of ¼ inch and comes in coils of 50, 75, 100, 150, and 200 feet. One hundred feet of fish wire weighs 2 pounds, 8 ounces.

If you are working with vertical pipe, you can fish with lengths of small chain. You insert the chain snake at the top and gravity insures that it descends in a straight line. The chain jingles as it passes down the pipe and lets you follow its progress inside the walls.

When the end of the chain arrives at the exit, you attach your conductor to it. You then pull your conductor up through the pipe. This is the easiest of the fishing techniques.

FISHING TECHNIQUE WITH ONE TAPE

When you obtain your reel of fish wire, you'll have to fashion a hook on the end. The manufacturer doesn't provide this feature.

Heat the end of the fish tape red hot to get rid of the tempering and then bend it into a hook. Let it cool slowly so the metal doesn't become brittle.

The fish tape's hook is bent as shown in the illustrations. The hook should be formed so that it will slide smoothly past any small obstructions without getting hung up.

During an actual run, you simply feed the hooked end into the conduit. Keep feeding until the hook emerges from the other end. Then firmly attach the conductor to the hook as shown in the illustrations (Figs. 6-6 and 6-7).

Withdraw the snake slowly until the hook is returned to your hands. It sounds easy, and often it is. At times, though, you'll hit snags in the pipe, especially other wires. Patiently you must work the snake loose until the wire returns to your hands.

FISHING TECHNIQUE WITH TWO TAPES

Most of the time one fish tape works well in pipe, but on occasion you will need two tapes. The length of pipe might be too long or the pull is too difficult for one tape.

The second tape is pushed in from the other end. You try to join the two hooks. You'll hear or feel them touch together (Fig. 6-8).

Taking note of the orientation of each hook, you work the two snakes till the hooks engage. When they do, you can withdraw one or the other according to your purpose.

When you are pulling wire—through walls, around corners, etc.—without the benefit of conduit, you will have a strong need for two tapes. Especially behind closed walls.

For example, suppose you want to pull a wire for a ceiling outlet from a wall switch. Your chances are slim to feed a #14 wire from the switch up the wall and across the ceiling to the outlet.

With two snakes the operation is easy. You feed one snake into the outlet hole and the other into the switch hole. The two hooks approach each other. You continue to advance the snakes until they meet.

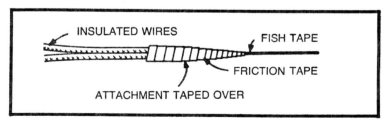

Fig. 6-7. To smooth out the wire wrapping on a hook, it's good technique to tape over the attachment.

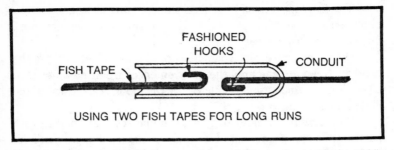

FASHIONED HOOKS

FISH TAPE

CONDUIT

USING TWO FISH TAPES FOR LONG RUNS

Fig. 6-8. In long runs two fish tapes can be used. The hooks meet in the middle and are engaged.

Then by feel you get the two hooks to engage. Lastly you pull one of the snakes all the way through. Now you are ready to pull the wire with the aid of the snake.

PULLING-IN LINE

For #14 and #12 wires, the fishing tape is more than adequate. But when you work with the larger sizes of wires, the fishing tape is not strong enough. It is bad technique to try to pull a #8 wire with a ¼-inch fish tape. What you need is a snake called a "pulling-in line."

You could even use a pulling-in line with #14 and #12 wire if its insulation is not smooth or if the pull has a lot of bends or if the conduit already contains a complement of wires.

A pulling-in line can be a length of ordinary rope (Fig. 6-9), sash cord, nylon rope or galvanized steel wire. The technique requires that the regular fish tape go first. The pulling-in line is then attached to the fish tape and the conductor to the pulling-in line.

If it seems like a lot of work, it is. Yet it's the only right way to get some tough pulling jobs done.

Great care should be taken when you attach one part of the snake to another.

PUTTING THE SNAKE TOGETHER

When you attach a conductor to the hook on a fish tape, you have a single temporary attachment. When you use a pulling-in wire, you have two temporary attachments. The first attachment is between the first tape and the pulling-in wire and the second is between the pulling-in wire and the conductor.

Some pulls are easy and your own arm strength is enough for working the conductor through the conduit. Other pulls are difficult and can require up to 1000 pounds of force. Residential pulls normally aren't overly difficult, but commercial and industrial wiring jobs present many hard pulls.

There are ways, however, to accomplish even the most difficult pulls. When your personal strength isn't sufficient, there are pulling devices available ranging from hand-operated winches to powerful motorized equipment.

Whatever the pull, your temporary attachments must hold. It's a costly situation when one of your attachments separates in the middle of a pull.

Good technique will have you bare the insulation on the end of the conductor for a distance of six inches. The conductor is fastened to the hook section of the wire in any of several ways.

The fastening is smoothed out by covering it with friction tape. If there are two or more conductors, you wrap them around the hook as you did the single conductor. This works well with easy pulls that do not require the use of a pulling-in wire.

If the pull is difficult, a pulling-in wire will be required and the snake parts will have to be fastened more rigorously.

It is convenient to have a pulling-in wire with a few small chain links joined to the end. The chain cannot be any larger in diameter than the line. The links are handy to attach to and will only slip loose reluctantly.

If more than one conductor is being pulled at the same time, it is a good idea to solder the attachment as an additional precaution after the customary wrapping of the conductors around the hook or chain ends. Then apply friction tape or electrician's plastic tape over the conductor and solder.

Another useful tool for pulling is the metal basket grip (Fig. 6-10). It is available in many sizes. The weblike grip slips over the ends of the conductors and holds tight. As you pull, the grip gets

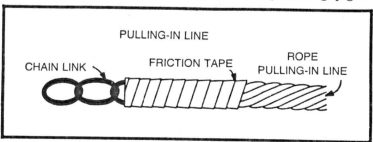

Fig. 6-9. For heavy pulls, in addition to the fish tape a pulling-in line can be used.

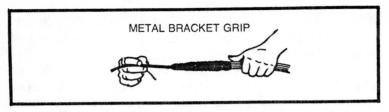

METAL BRACKET GRIP

Fig. 6-10. Electricians use metal basket grips with pulling-in lines to hold tight to the conductors.

tighter. The harder the pull, the tighter the grip becomes. The basket grip has a maximum usefulness of 1000 pounds.

USEFUL TIPS FOR PULLING

It is wrong to pull conductors into a conduit with so much force that they can't be removed ever afterward. One of the principal reasons for using conduit is to facilitate the future removal of conductors. The conduit is there to enable you to work behind walls or under concrete slabs without opening the walls or chopping through the slab. You should originally use a pipe that is large enough to handle the conductors with plenty of room to spare.

That is why the Code recommends filling the conduit to only 40 percent of capacity in many applications.

As you feed conductors into conduit, take care that they slide straight and do not twist or bunch up. Make sure that the conductors do not cross over each other. Ideally, each should lie straight throughout the pipe. This is scarcely attainable, but it is to be strived for.

Make certain the inside of the pipe is clean and dry. If you imitate the practice of some workmen who stuff the gaps in the ends of the conduits as a precaution against the accidental entry of foreign matter, make sure that you don't use paper as your stuffing material.

Before pulling the conductors, lubricate the inside of the pipe. Blow some powdered soapstone into the pipe, especially around bends and elbows. Rub some of the soapstone on the conductors as you feed them into the pipe.

A special liquid lubricant called Wire Lube is very useful for pulling wire. You can spread the pastelike material on the conductor. The lubricant dries into a slippery and long-lived powder which will make all future conductor feedings and withdrawings easier. Wire Lube also contains an anti-corrosion additive that contributes to the durability of the insulation.

TENSION IN POUNDS

To calculate the tension in pounds of a pull is easy. You can calculate the tension of a straight ductwork pull with the formula

$$T = LWF$$

T equals tension in pounds, L is the length of the conduit in feet, W is the weight of the cable in pounds, and F is the coefficient of friction (around 0.5).

Suppose you are pulling 5 pounds of cable through 100 feet of pipe.

$$T = 100 \times 5 \times .5 = 250 \text{ lb. of tension}$$

A pull of 250 pounds is a tough one and perhaps is a bit more than can be done easily. Another man to help you pull or a hand operated winch would be a big help in getting the conductors in place.

If the conduit were only 25 feet long:

$$T = 25 \times 5 \times .5 = 62.5 \text{ lbs. of tension}$$

PULLING CONDUCTORS WITHOUT CONDUIT

In the home, especially when you add electrical systems to existent ones, you'll often not use pipe. You'll be pulling wire through walls without the benefit of tubing.

Usually, strength is not a factor. Most of the friction you'll run into will be caused by contact with the insulation pads in the wall or with the perimeter of the holes themselves. If a roll of cable weighs ten pounds, the pulling effort is negligible. It needn't even be calculated.

There are excellent non-conduit types of wire like BX and Romex that, once in place and installed correctly, are nearly as rigid and durable as conduit.

This kind of cable can be run and fished easily through partitions, studs, under wooden floors, above ceilings, around door frames, through attics, etc. It can be buried underground or run through the air holes in masonry block walls with no possibility of moisture damage.

FITTINGS

If you look in Article 100 of the Code book, you'll find separate definitions for "fittings" and "devices." These terms refer to two different kinds of electrical materials, but there is confusion about the distinctions between them.

FITTINGS VERSUS DEVICES

The Code says that a fitting is "an accessory such as a locknut, bushing or other part of a wiring system which is intended primarily to perform a mechanical rather than an electrical function."

On the other hand, the Code defines a device as "a unit of an electrical system which is intended to carry but not utilize electric energy." Devices are discussed in Chapter 8, but it should be apparent now that a device has current pass through it. Devices are typically switches, fuses, etc.

A fitting does not have current pass through it. It is in the system to mechanically hold the wiring in place. It lies outside the wire and insulation. It should be grounded for safety's sake, even though it does not carry current in normal operation.

Let us emphasize the functional difference between the two electrical materials. A device joins the electron flow of one conductor to another and thus passes current from place to place. A fitting fastens together two conductors physically so they won't separate. A fitting is grounded where necessary but does not pass current.

Typical fittings are displayed in the accompanying illustrations (Fig. 7-1).

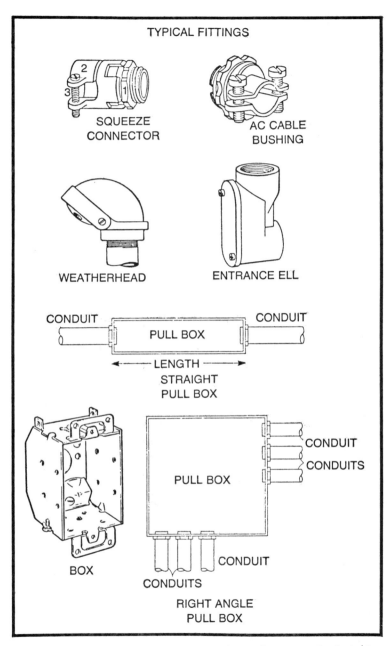

TYPICAL FITTINGS

SQUEEZE
CONNECTOR

AC CABLE
BUSHING

WEATHERHEAD

ENTRANCE ELL

CONDUIT   CONDUIT

PULL BOX

←——— LENGTH ———→
STRAIGHT
PULL BOX

CONDUIT

CONDUITS

PULL BOX

CONDUIT

BOX

CONDUITS

RIGHT ANGLE
PULL BOX

Fig. 7-1. A fitting is a part of the wiring system that performs a mechanical rather than an electrical function.

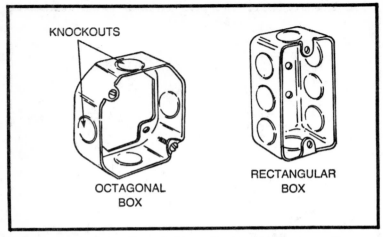

KNOCKOUTS

OCTAGONAL
BOX

RECTANGULAR
BOX

Fig. 7-2. Boxes are manufactured with machine punched knockouts. These are easily removed to run cable or conduit.

BOXES

Boxes do not pass current. They are a type of fitting. They are sometimes called devices but by Code definition they are not. They perform the job of safely containing devices.

Until electrical power on a large scale was introduced, boxes were not used much. Electricians ran wires inside walls or along the outer surface of walls right up to the switches and fuses. The wires were connected directly to the devices.

Fixtures were hung directly from ceilings and unhoused switches were notched into the wall plaster. Yes, this was a hazardous way to electrify a house, and the homeowner had to be careful. Fires and electric shocks often occurred.

Over the years a safe housing for connections evolved in the form of outlet boxes. Instead of connections inside the walls and unavoidable sparking near inflammable materials, the connections all went safely into the boxes and the earlier hazards were largely eliminated.

In addition, modern boxes are all grounded, adding to the safety factor. Boxes come in four general shapes. They are octagonal, rectangular (Fig. 7-2), square, and circular, with variations on each general shape.

The Code requires all boxes to be supported carefully. There are Code standards which specify mechanical strength and methods for the installation of boxes in various places such as floors, wet locations, and corrosive places.

BOXES IN COMMON

Boxes are called "outlet boxes." They usually have "knockouts" spaced conveniently over them (see Fig. 7-20). The knockouts are machine-punched circles cut in (but not removed from) the metal box. You can easily punch out the knockout to obtain an opening for running a cable or for attaching a pipe to the box.

Boxes are made of steel or plastic. The sheet steel boxes are built in steel gage thicknesses of #10 through #14. The steel is galvanized; this means that the steel is coated with zinc. The steel boxes can be used either with pipe or cable.

Plastic boxes come in all sizes but cannot be used with pipe. The Code permits them to be used with cable only. The cable must have a grounding conductor and the plastic box must have a continuous ground run through it, even though the plastic itself is an insulator.

Steel boxes are required to have a good solid electrical connection to the grounded pipe. Without exception a box should be continuous to ground. The connections must be strong enough to withstand the ravages of time, corrosion, structural settling, and other physical stresses.

KNOCKOUT TECHNIQUE

If you look at the knockouts on a box, you'll see that the metal is completely cut around the circumference of the circle except at one spot. This intact point is called the anchor.

To remove a knockout, you hit it with the end of a heavy screwdriver. A small hammer or a heavy pair of pliers would also provide a strong enough blow to bend open the knockout. Then, if you grasp the knockout disc with the pliers, you can twist it off easily.

Some boxes have a pryout type of knockout. The pryout type features a small slot in the center of the knockout disc. A screwdriver goes into the slot and you pry out the disc.

Either way, the Code only permits you to remove the knockout that you need to install the pipe or cable. If you accidentally take out a knockout, you must close up the hole or not use the box.

There are a number of plug type gadgets that enable you to close up any inadvertent holes in a box.

WHAT BOXES DO

Boxes used in interior wiring are classified as:

1—Outlet boxes
2—Utility boxes
3—Sectional switch boxes
4—Floor boxes

An outlet box is installed at any point in a wiring system where current can be obtained easily for supplying lamps or appliances.

The utility box is like an outlet box but is designed to handle exposed wiring.

The sectional switch box is used in concealed wiring when not buried in masonry.

Floor boxes are used to make outlets in floors and to provide access to conduit wiring.

You must know your boxes so you can choose the correct box for each job. There are hundreds of variations. You can safely interchange one type of box for another. However, just as straight runs and square bends of pipe are the mark of the expert, the use of the appropriate box shows that a skilled worker has performed the installation.

The appropriate box will quickly attach to the supporting structure. Wooden studs require one kind of box while metal studs require another.

The right box will have enough space inside to handle the number of wires that will occupy it. The Code is very strict about the number of conductors placed in a box. Let's examine some of the box considerations.

BOX SUPPORTS

The commonest method to support an outlet box is to use a hanger (Fig. 7-3). The hangers and boxes are made to be used together. To attach the hanger to the box, you remove the knockout disc in the center of the box's bottom plate. The box is then slipped over the fixture stud that is part of the hanger. Then a locknut on the inside of the box is tightened onto the projecting fixture stud. The 1½-inch-thick box is recessed so that its outer edge lies flush with the plaster.

Pay attention to the fixture stud. It takes up room inside the box. Since it does take up room, the Code requires that you include it when you calculate the volume of the conductors in the box.

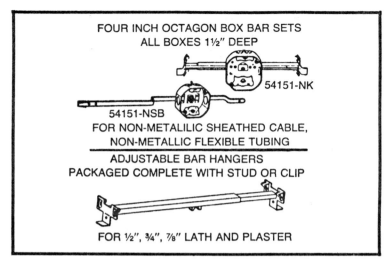

FOUR INCH OCTAGON BOX BAR SETS
ALL BOXES 1½″ DEEP

54151-NK

54151-NSB

FOR NON-METALILIC SHEATHED CABLE,
NON-METALLIC FLEXIBLE TUBING

ADJUSTABLE BAR HANGERS
PACKAGED COMPLETE WITH STUD OR CLIP

FOR ½″, ¾″, ⅞″ LATH AND PLASTER

Fig. 7-3. The fixture stud is part of the hanger the box is attached to in the ceiling or wall.

For example, you could count the fixture stud as one #8 wire. One #8 wire occupies 3 cubic inches of space in the box. We'll develop this subject further in later sections of the chapter.

In addition to fastening the box to the hanger, the fixture stud supports the fixture that is joined to the box.

Boxes can be purchased as individual units or as box-hanger assemblies. The hanger is hung between two joists. The hole for the box is cut exactly and no open space is permitted around the box.

This type of box support is used with cable conductors. It is not used with pipe. A deeper hanger is needed with conduit installations so that the offset bend can pass straight into the box.

This places the box about ½ inch lower than the surface of the plaster. The box, though, is supposed to be flush. The way to make it flush is to use a box cover. There are many styles of covers to suit the various applications of the boxes (Fig. 7-4).

There are covers for switches, receptacles, and so on. The covers are mainly used in conduit work.

You might think that a hanger isn't necessary since a box is firmly supported by its attachment to the conduit. However, the Code writers deem this arrangement not sturdy enough. Because the box gets manual handling, it requires additional, direct support.

Another common kind of box support is the bracket. A bracket is an ear type of thing that is attached to the box during manufacture.

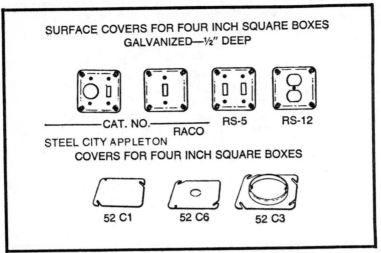

Fig. 7-4. Box covers are used to seal off and protect the electrical connections inside the box.

Some brackets require nails, others screws, and others bolts. The type of fastener used depends on the composition of the support to which the bracket will be attached.

Some boxes are affixed to mounting strips of wood or steel. The illustrations show a number of different arrangements.

BOX COVERS

Outlet boxes must be covered. The cover is a special box cover, perhaps for a box from which a fixture is hung. A cover must be used.

Quite often a box is installed in the middle of a conduit run. It is there to give access to the conductors. It may be used to pull wires or to form a junction. In these instances it would be a "pull box" or a "junction box." Pull boxes get blank covers.

Some blank covers have knockouts in them. If so, the knockouts will most likely lie in the center of the cover. This is a specialty cover that often is used to accommodate the armored cables which connect to large loads (a motor, perhaps).

BOX SHAPE DETAILS

It was mentioned previously that boxes are made in four shapes: round, octagonal, square (Fig. 7-5) and oblong. Round and octagonal boxes are used mostly in ceiling outlets for concealed wiring. Octagonal boxes are also used in wall-bracket lighting outlets, but round boxes are not. Both styles are available with fixture studs of ⅜-inch or ½-inch diameter or simply with a hole ready to receive a fixture stud.

Round boxes can only be attached to at the flat bottom. You cannot attach to the side of a round box because there are no flat surfaces.

You cannot get a good solid physical and electrical connection with a locknut or bushing on the sides of a round box. If you try to tighten a nut against the outer surface, the center of the nut will sit flush enough but the edges will have a spacing between them and the side of the box. Trying to tighten a nut against the inner surface is equally futile.

Square and oblong boxes are used mainly on masonry, brick or tile walls. Square boxes are used occasionally as ceiling outlets. When they are installed in the ceiling, the fixture stud is used as it is with the round and octagonal boxes.

Oblong metal boxes are frequently called "gangboxes." There are more cubic inches inside an oblong box than inside a square box. It is called a gangbox since it can handle a number of devices ganged together. For example, a three-gang box can handle three switches side by side. Don't get the words oblong and rectangular confused. They mean the same thing.

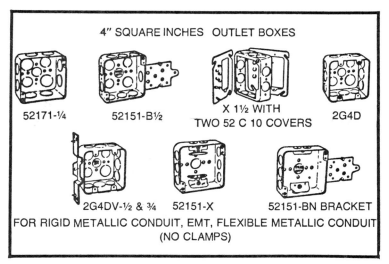

Fig. 7-5. Boxes, boxes and more boxes, each suited to a particular use.

Box Dimension, Inches Trade Size or Type	Min. Cu. In. Cap.	Maximum Number of Conductors				
		No. 14.	No. 12	No. 10	No. 8	No. 6
4 × 1¼ Round or Octagonal	12.5	6	5	5	4	0
4 × 1½ Round or Octagonal	15.5	7	6	6	5	0
4 × 2⅛ Round or Octagonal	21.5	10	9	8	7	0
4 × 1¼ Square	18.0	9	8	7	6	0
4 × 1½ Square	21.0	10	9	8	7	0
4 × 2⅛ Square	30.3	15	13	12	10	6*
4 11/16 × 1¼ Square	25.5	12	11	10	8	0
4 11/16 × 1½ Squre	29.5	14	13	11	9	0
4 11/16 × 2⅛ Square	42.0	21	18	16	14	6
3 × 2 × 1½ Device	7.5	3	3	3	2	0
3 × 2 × 2 Device	10.0	5	4	4	3	0
3 × 2 × 2¼ Device	10.5	5	4	4	3	0
3 × 2 × 2½ Device	12.5	6	5	5	4	0
3 × 2 × 2¾ Device	14.0	7	6	5	4	0
3 × 2 × 3½ Device	18.0	9	8	7	6	0
4 × 2⅛ × 1½ Device	10.3	5	4	4	3	0
4 × 2⅛ × 1⅞ Device	13.0	6	5	5	4	0
4 × 2⅛ × 2⅛ Device	14.5	7	6	5	4	0
3¾ × 2 × 2½ Masonry Box/Gang	14.0	7	6	5	4	0
3¾ × 2 × 3½ Masonry Box/Gang	21.0	10	9	8	7	0
FS—Minimum Internal Depth 1¾ Single Cover/Gang	13.5	6	6	5	4	0
FD—Minimum Internal Depth 2⅜ Singl Cover/Gang	18.0	9	8	7	6	3
FS—Minimum Internal Depth 1¾ Multiple Cover/Gang	18.0	9	8	7	6	0
FD—Minimum Internal Depth 2⅜ Multiple Cover/Gang	24.0	12	10	9	8	4

*Note to be used as a pull box. For termination only.

Fig. 7-6(a). Table 370-6 (a) lists all the box trade sizes and the cubic-inch capacity of each.

Volume Required Per Conductor

Size of Conductor	Free Space Within Box For Each Conductor
No. 14	2. cubic inches
No. 12	2.25 cubic inches
No. 10	2.5 cubic inches
No. 8	3. cubic inches
No. 6	5. cubic inches

Fig. 7-6(b). Table 370-6 (b) provides the calculated cubic inch space each conductor takes up in a box.

METAL BOXES

If you look at Fig. 7-6(a), you'll see that the first column lists boxes of many sizes, shapes and styles. The boxes have dimensions. The Code specifies that "no box shall have an internal depth of less than 1/2 inch."

The boxes in the table have a minimum depth of 1¼ inches and a maximum depth of 3½ inches. A size of around 2½ inches is most often used.

The table's second column shows the capacity of the boxes in cubic inches. The capacity of a box is important since it reveals how many wires can be put into a box. The amount of space occupied by wires of every size has been calculated. The Code is very precise about the cubic-inch displacement of each wire.

Figure 7-6(b) records the amount of free space that should surround conductors of various sizes. The sizes of the conductors are correlated with the required free space as expressed in cubic inches. For example, a #8 wire requires three cubic inches of free space around it.

The wires listed are #14, #12, #10, #8, and #6. Larger wires have to conform with the specifications of another table in the Code book.

Referring again to Fig. 7-6(a), observe that the wires are listed by size. The list shows the number of identical wires you can run into a box. For example, in the 4-by 1¼-inch round box with a capacity of 12.5 cubic inches you can use six #14 wires, five #12 wires, five #10 wires or four #8 wires. You are not permitted to put any #6 wires into a box of that size.

That list is fine if you are running, for example, all #12 wires. When you are, you can use the column with the #12 conductors shown for each type of box.

WHEN DIFFERENT SIZE WIRES ARE RUN

What about the cases where you must run wires of a number of different sizes into a box? The situation becomes a bit more complicated.

That is because the volume of the box has to be compared with the free space required by each conductor.

A typical case would occur when you had to choose a box to hold three #12 wires, three #10's, two #8's and two #12 ground wires. There'll also be a fixture stud to contend with. (See Fig. 7-7.)

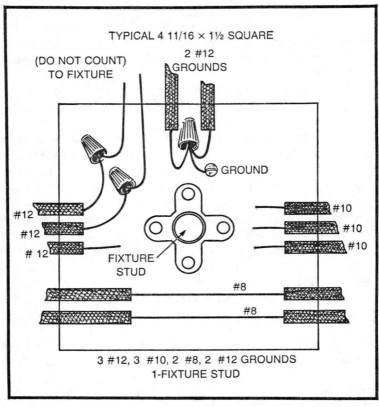

TYPICAL 4 11/16 × 1½ SQUARE

(DO NOT COUNT)
TO FIXTURE

2 #12
GROUNDS

GROUND

#12
#12
12

FIXTURE
STUD

#10
#10
#10

#8

#8

3 #12, 3 #10, 2 #8, 2 #12 GROUNDS
1-FIXTURE STUD

Fig. 7-7. First the cubic inch requirement is calculated, then the box size is chosen from the list.

You should proceed by calculating the number of cubic inches these wires and the fixture stud require. You can obtain the requirement by consulting Fig. 7-6(b).

3 #12 is 3 × 2.25 cu. in.	= 6.75
3 #10 is 3 × 2.5 cu. in.	= 7.50
2 #8 is 2 × 3 cu. in.	= 6.00
One fixture stud is 1 × 3 cu. in.	= 3.00
2 #12 ground wires is 1 × 2.25 cu. in.	= 2.25
TOTAL cu. in.	= 25.50

If you look at Fig. 7-6(b), you'll find an entry of 25.5 cu. in. It is a 4 $^{11}/_{16}$ × 1¼ square box. You can choose that box. Or you can choose a box slightly larger, such as the 4$^{11}/_{16}$ × 1½ square box.

Looking back at the calculations, note that a fixture stud occupies the same amount of space as the largest size wire (#8). Figure 7-6(b) says that is a displacement of 3 cubic inches. It states this fact in a roundabout way: "Where one or more fixture studs are in the box, the number of conductors shall be one less than shown in the Tables."

Also note that the two #12 ground wires need only as much space as one wire. Elsewhere, the Code book in its characteristic roundabout fashion again refers to the subject: "A further deduction of one conductor shall be made for *one or more* grounding conductors entering the box."

The Code language is sometimes puzzling and indirect, but you must learn the various things that are meant.

NONMETALLIC BOXES

When a box is attached to metal conduit, a good electrical connection is required. Any break in the ground line is hazardous. (You'll learn why later in the book.) Because of this, you are not permitted to use a nonmetallic box to house the connections of your conductors.

A nonmetallic wiring run has no provision for exterior grounding. It is insulated in entirety. The grounding of nonmetallic cable and nonmetallic conduit is accomplished in the interior by a separate conductor which acts as a continuous ground.

Nonmetallic boxes are constructed of nonconductive materials and thus are useful for insulating purposes. They are variously made from porcelain, Bakelite and polyvinyl chloride (PVC). In fact, nonmetallic conduit is often called PVC conduit, in reference to its composition.

Porcelain boxes are used with nonmetallic cable but not with conduit-contained wiring. They are easily cracked or broken and should not be installed in places where they are liable to be bumped or struck.

The most popular nonmetallic box is the variety made from Bakelite. It's the familiar brown plastic box. It is rugged and comes in many convenient sizes. Fiber-reinforced construction gives it extraordinary strength.

PVC boxes were originally created to be used with rigid PVC conduit. PVC is chemical-resistant and permanently watertight. It is a good material to use in wet, chemically hazardous areas. The

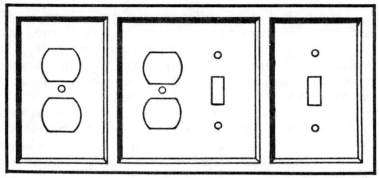

Fig. 7-8. Flush plates are not box covers. They are used mostly to decorate and finish a job.

nonmagnetic property of the PVC box (and all other nonmetallic boxes) is useful in industrial facilities where large magnetic fields are present.

FLUSH PLATES

Flush plates (Fig. 7-8) are used on the fronts of boxes. They are NOT box covers. They are not even the same kind of hardware.

A box cover is manufactured in many styles and its purpose is to seal off the interior of the box from easy access. A flush plate is only a surface covering used to conceal the wiring inside the box for a neater appearance. The plate is drawn flush to the wall by screws passing through the plate and threading into tapped holes in the box. You are quite familiar with the arrangement. If you look around the room you'll see some of these plates in action. They come in plain or fancy decorative finishes. No doubt you have handled one.

The plates have no electrical function. Some are nonmetallic, others are metallic. They are available in Bakelite, stainless steel, enameled metal, brass, etc.

BETWEEN THE PIPE AND THE BOX

An important fitting, and one which must be installed carefully, is the locknut-bushing system (Fig. 7-9) used on conduit ends. It mechanically secures the conduit to the box and produces a continuous ground.

The locknut and bushing are also effective at closing gaps. Wherever a conductor enters a box, it must do so in the safe confines of the cable insulation or pipe. The locknut-bushing makes an excellent seal.

A locknut-bushing connection is made like this. First, the locknut goes onto the conduit. It is turned halfway along the threads of the rigid metal conduit. The tip of the conduit is then slipped through the knockout into the box and the bushing is turned onto the threads from the inside. The bushing is turned down as snugly as possible. Then the locknut is tightened with a wrench. The tightening of the locknut draws the bushing close against the box. The locknut bites into the metal of the box and makes a good solid ground.

When cable or EMT is used, a cable connector instead of the locknut-bushing arrangement is needed. The connector is screwed tightly to the end of the cable or EMT by tightening the two screws. The connector is then slipped through the knockout and a locknut is installed inside the box. As before, the locknut is tightened down until it bites into the metal.

CONDUIT FITTINGS

Conduit fittings are handy gadgets that let you adapt pipe in tough mechanical situations. The illustrations show a variety of fittings (Fig. 7-10). Note the ½-inch offset fitting. It saves you the trouble of making awkward offset bends. Instead of bending the pipe, you simply secure an offset fitting to the box.

A 90-degree elbow is also available for your use. All you do is attach the two ends of pipe to the elbow. No pipe shaping is needed and you have a good 90-degree bend.

Some of the 90-degree fittings have covers that give access to the interior. You can use the cover opening for pulling wires or for inspecting the integrity of the run.

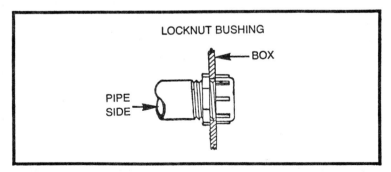

Fig. 7-9. The locknut bushing connects the conduit to a box and maintains the continuity of the ground.

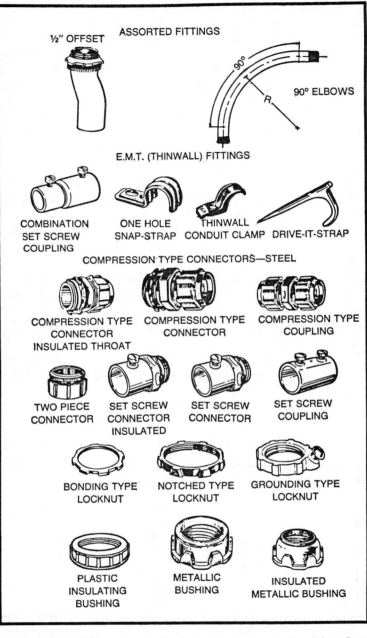

ASSORTED FITTINGS

½" OFFSET

90°

90° ELBOWS

R

E.M.T. (THINWALL) FITTINGS

COMBINATION
SET SCREW
COUPLING

ONE HOLE
SNAP-STRAP

THINWALL
CONDUIT CLAMP

DRIVE-IT-STRAP

COMPRESSION TYPE CONNECTORS—STEEL

COMPRESSION TYPE
CONNECTOR
INSULATED THROAT

COMPRESSION TYPE
CONNECTOR

COMPRESSION TYPE
COUPLING

TWO PIECE
CONNECTOR

SET SCREW
CONNECTOR
INSULATED

SET SCREW
CONNECTOR

SET SCREW
COUPLING

BONDING TYPE
LOCKNUT

NOTCHED TYPE
LOCKNUT

GROUNDING TYPE
LOCKNUT

PLASTIC
INSULATING
BUSHING

METALLIC
BUSHING

INSULATED
METALLIC BUSHING

Fig. 7-10. Offset fittings, pipe bends, entrance ells, threaded and threadless fittings are all available.

158

Entrance ells (elbows) connect service conduit to the spot where the house circuit joins the utility company's entrance line. An entrance ell is often used where the line passes through a wall.

The conduit fittings used on rigid metal pipe have threads. Fittings for EMT are threadless. Springs, setscrews and clamps take the place of threads. EMT fittings are less expensive than rigid metal conduit fittings.

If the threadless fittings are installed with care, they are excellent for use in wet locations or for burial in masonry and fill dirt. They are as waterproof as their rigid metal counterparts.

PULL BOXES

Since it is not unusual to need 1000 pounds of pull to get wires through a conduit, a little help is always welcome.

In the wire pulling formula, one of the factors is L, the length of the run. If you can reduce L, you can reduce the amount of pounds needed to pull the wire.

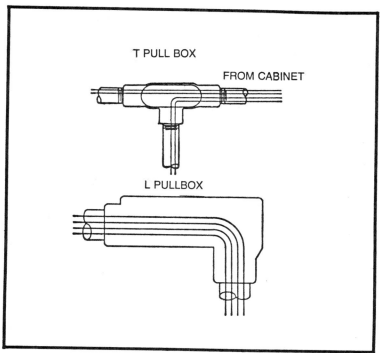

Fig. 7-11. A pull box is essential during wire pulls to reduce the length of a run. The wire can be pulled to a pull box that is located midway in the run.

As mentioned, a pull box (Fig. 7-11) can be substituted for an elbow. But a pull box can also be straight, and, if so, can provide a "manhole" somewhere along the conduit run which will be helpful for pulling the wire.

Pull boxes are usually made of sheet steel and are available in many sizes. It is easy to make 45-degree, 90-degree and U-turns with the aid of pull boxes.

In residential jobs, a pull box usually has one entrance and one exit for the wire. Only two knockouts are removed from the box.

In larger jobs, many knockouts might be removed and many conduits installed. When a single pull box is fitted with a number of conduit runs, the Code gets complicated about the subject.

It is not too hard to understand the requirements for straight pulls (where there is to be no bend in the wiring). The Code says that for straight pulls the length of the box must be at least 8 times the trade diameter of the largest conduit entering the box.

For example, suppose the largest conduit entering the box is a 2-inch pipe. (The Code says trade diameter, not actual diameter.) Then the pull box length must be at least 16 inches.

It's always best to get a slightly larger pull box than required, because it's easy to crowd a box to the point where you can't get purchase for pulling.

ANGLE PULLS

When the box is used for an angle pull, the wording of the Code gets confusing. When only one conduit or cable is involved, the meaning is fairly clear: the distance to the wall opposite the wire entrance must be at least 6 times the diameter of the conduit.

For example, a ¾-inch conduit entering a pull box for a 90-degree pull has a proper distance to the opposite wall of $6 \times ¾ = 18/4 = 4½$ inches. This should provide sufficient space to pull and bend the wire.

If more than a single conduit is involved, the calculation becomes complex. The distance to the opposite wall must be a quantity which is 6 times the diameter of the largest conduit added to the sum of the diameters of all the other conduits entering the pull box.

Further, the Code prescribes that the closest distance between an entrance conduit and an exit conduit must be at least 6 times the diameter of the conduit. If the entrance conduit differs in size from the exit conduit, use the larger size for the calculation.

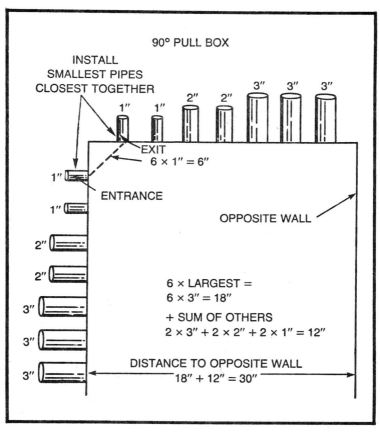

Fig. 7-12. In this example, the Code leads the way to a calculation. A 30-inch pull box is needed.

A typical 90-degree pull box on a large job might handle three 3-inch pipes, two 2-inch pipes and two 1-inch pipes (Fig. 7-12). You would install them so the wires can leave in the same order they entered the box. What must the dimensions of the box be?

First of all, it is good technique to install the pipes so that the pipes of the smallest diameter are arranged next to one another on the inside of the bend as in Fig. 7-12.

If the 1-inch pipes lie next to one another, then the distance from center to center of the innermost entrance and exit points must be 6 × 1 = 6 inches. That takes care of that requirement.

Next is the opposite wall dimension. According to the Code a distance is needed which equals 6 times the diameter (3 inches) of

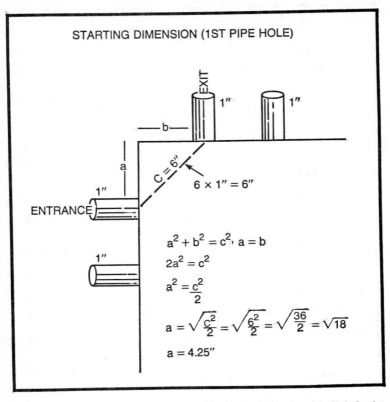

STARTING DIMENSION (1ST PIPE HOLE)

$$a^2 + b^2 = c^2, a = b$$
$$2a^2 = c^2$$
$$a^2 = \frac{c^2}{2}$$
$$a = \sqrt{\frac{c^2}{2}} = \sqrt{\frac{6^2}{2}} = \sqrt{\frac{36}{2}} = \sqrt{18}$$
$$a = 4.25''$$

Fig. 7-13. The distance from the corner of the box to the center of the hole for the 1st pipe works out to 4¼".

the largest pipe added to the sum of the diameters of all the other pipes.

$$
\begin{array}{rl}
6 \times 3'' = & 18'' \\
2 \times 3'' = & 6'' \\
2 \times 2'' = & 4'' \\
2 \times 1'' = & 2'' \\
\hline
= & 30''
\end{array}
$$
(distance to opposite wall)

GETTING YOUR STARTING DIMENSION

In actual practice, when you prepare to make holes for the first 1-inch pipe, you can measure off the distance in a couple of ways.

One, take a rule and span 6 inches from center to center for the 1-inch pipe holes. That is the easy way (Fig. 7-12).

A second way is to use simple algebra (Fig. 7-13). If the 6-inch span is considered the hypotenuse of a right triangle, then the distances from the point of the box to the centers of the holes are the sides of the triangle. This right triangle can be solved with the formula $a^2 + b^2 + c^2$. Since a^2 and b^2 are of the same length, they can be expressed as $2 a^2$. C is 6 inches.

$$2 a^2 = c^2$$
$$2 a^2 = (6)^2$$
$$2 a^2 = 36$$
$$a^2 = 36/2 = 18$$
$$a = \sqrt{18}$$
$$a = 4.25 \text{ inches}$$

In conclusion, for this installation you need a pull box 30 inches square and the distance from the corner of the box to the center of the first hole is 4.25 inches.

This box meets the minimum Code requirements. It is better to use a slightly larger box and avoid crowding.

Devices

At the beginning of the last chapter we mentioned the confusion that many people have about the distinctions between fittings and devices. Now that you have a firm understanding of what fittings are, let us delve into the subject of devices.

NEC DEFINITION OF A DEVICE

Article 100 of the Code says, "Device—a unit of an electrical system which is intended to carry but not utilize energy."

If you consider what an energy utilizer is, perplexity can develop. A lamp or appliance utilizes energy. Yet, at first glance, a lamp or appliance would appear to be a device. Not according to Code. The Code has other ideas.

You might also argue with the framers of the Code about whether a wire is a device since it carriers but does not utilize energy. Again confusion can occur, because according to the Code wire is not a device but a conductor.

The electrician has to consider a device as a unit of the system that stands between the wire and the lamp or appliance. There are six general types of devices:

        1—Switches
        2—Fuses
        3—Circuit breakers
        4—Controllers
        5—Receptacles
        6—Lampholders

That is the roundup of devices. The switches, fuses and circuit breakers are installed in the hot line to disconnect the line when the need arises.

A controller is installed in the hot line to adjust the amount of current that can pass from the conductor to the energy utilizer.

Receptacles and lampholders connect to the hot and neutral lines so that lamps and appliances can be energized by either plugging into or attaching to the circuit.

SWITCHES AND CIRCUIT BREAKERS

You've seen switches on the wall and circuit breakers and fuses in your house panel. In the past you've purchased and even installed some of these items.

They are familiar objects, but don't let their familiarity lull you into believing you know all about them. There is still plenty to learn.

A switch is defined as a device for making, breaking, or changing connections in an electrical circuit. The first fact to learn is that the switch is connected in series in the hot line. It is never connected into the neutral line.

A switch is designed to handle a certain number of amperes. Common switches are 15- or 10-amp types. They have two terminals. To install one, you open up a 120-volt line and connect the ends of the broken hot line to the two terminals of the switch (Fig. 8-1a).

A switch is manually operated. You turn the switch off and on, which means you are opening the hot line (off) or shorting the two ends of the hot line back together (on). In other words, you control the mode of operation. Should a lamp be on or off? It is your decision.

A circuit breaker is a switching device, but it is not a switch in the narrow sense of the word. A circuit breaker, though, is also installed in series in the hot line. Like an ordinary switch, it is designed to handle a certain number of amps. Typical circuit breakers handle 15, 20, 30, 40 and 50 amps. Circuit breakers also have two terminals which when installed close a hot line. However, a circuit breaker's normal function isn't to serve just another manually operated switch.

While you can flick the toggle of a circuit breaker to turn the current off or on, its real purpose is to monitor flow and to open automatically if a fault occurs which draws too much current through the hot line (Fig. 8-1b).

When an unusually large quantity of current flows, the breaker, either due to heat or magnetism developed by the fault, snaps open. Automatic disconnect is the function of the breaker.

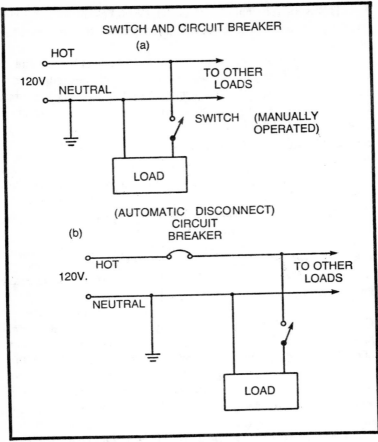

Fig. 8-1. Both a switch and circuit breaker are in series with the hot line, but perform different functions.

FUSES

The fuse was the predecessor of the circuit breaker. Panels have been called "fuse boxes" since the beginning of the wide use of electricity (Fig. 8-2).

The circuit breaker was devised as an improvement to fuses. When a fuse blows, it has to be replaced. When a circuit breaker blows, it only has to be reset. That saves time and the replacement cost.

A fuse is also a disconnect device like the breaker. You can turn off a circuit by unscrewing or unplugging a fuse. The circuit can be turned on again by re-installing the fuse.

This arrangement is used for small fuses such as the ones that protect branch circuits. The main fuse, however, has a companion device, the main switch, to turn off all the circuits at once. The same effect can be achieved by unscrewing the main fuse. Both the main switch and the main fuse can do the same job. The switch is easier and safer, though.

The best arrangement of all for cutting off every circuit at the same time is to use a panel that holds a fuse block for the main fuse. A fuse block is a plastic container that holds the main fuse system. The fuse block has a handle. When you want to disconnect the entire flow of electricity, you tug on the handle. The block pulls out and opens the main circuit. It is a combination form of main fuse and main switch (Fig. 8-3).

OVERCURRENT DEVICES

Fuses and circuit breakers are overcurrent devices. The Code defines a fuse as "an overcurrent device that is heated and severed by the passage of overcurrent through it."

The Code defines a circuit breaker as "a device designed to open and close a circuit by nonautomatic means, and to open the circuit automatically on a predetermined overload of current, without injury to itself when properly applied within its rating."

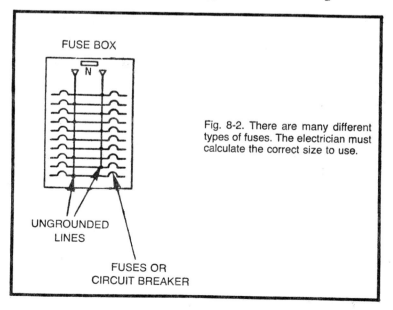

FUSE BOX

Fig. 8-2. There are many different types of fuses. The electrician must calculate the correct size to use.

UNGROUNDED LINES

FUSES OR CIRCUIT BREAKER

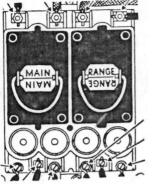

MAIN & RANGE
FUSE BLOCKS

MAIN

RANGE

Fig. 8-3. A fuse block is used to combine the jobs of the main switch and the main fuse.

In simpler terms, a fuse is a short length of metal ribbon made of an alloy that has a certain melting point. As long as a normal amount of current flows through it, the metal acts as a good conductor. As soon as an overcurrent begins to flow, the metal heats up and melts. This opens the circuit and any danger from the overcurrent is eliminated (Fig. 8-4).

The circuit breaker has an ON-OFF toggle switch. You can use the toggle to disconnect the circuit. The fuse has no such provision.

The circuit breaker also has a carefully calibrated bimetal strip. This is the same kind of bimetal strip found in thermostats. It is sensitive to any increase in temperature.

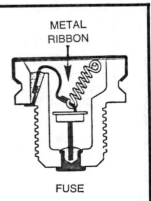

METAL
RIBBON

Fig. 8-4. A fuse has a short length of metal ribbon which heats up and melts when an excessive amount of current passes through it.

FUSE

If an overcurrent should occur, the bimetal strip heats up and bends. The bending trips the contacts and the ON position snaps to OFF (Fig. 8-5). This has the same effect as the melting of the fuse. Once the overcurrent fault is relieved, you can easily reset the toggle to ON.

Fuses and circuits breakers both are rated in amperes. In Article 240-6 of the Code the ampere ratings are listed. They are 15, 20, 25, 30, 35, 40, 45, 50, 60, 70, 80, 90, 100, 110, 125, 150, 175, 200, 225, 250, 300, 350, 400, 450, 500, 600, 700, 800, 1000, 1200, 1600, 2000, 2500, 3000, 4000, 5000, and 6000.

Fuse boxes are less expensive to install than circuit breaker panels. Yet the trend is away from fuses because of the superior merits of circuit breakers.

FIGURING OVERCURRENT DEVICE RATINGS

A 120-volt circuit has two lines. One is the neutral, which is grounded, and the other is the hot line, which is called "ungrounded." The neutral never gets an overcurrent device in this type of circuit. Usually it is grounded and at zero volts. If you touch it, there should be little or no shock. There is current in the neutral, the same amount as in the hot line, but there is no voltage pressure between you and the grounded neutral. You are both grounded at zero volts.

If you were to install an overcurrent device in the neutral, you might inadvertently discontinue the zero volts at that point. The neutral could become a hot line and shock you. The ground continuity must be maintained through the neutral wiring. More information about grounding appears in Chapter 10.

The overcurrent device is connected in series with each of the ungrounded conductors. The Code specifies this.

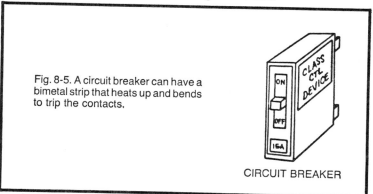

Fig. 8-5. A circuit breaker can have a bimetal strip that heats up and bends to trip the contacts.

CIRCUIT BREAKER

Each wire has a rating in amperes. This is the amperage capacity, called ampacity. It refers to the amount of current the wire can safely carry.

When you install an overcurrent device in series with a wire, the rating of the device is determined by the ampacity of the wire as prescribed by the Code. Whatever the wire can carry, the device must be able to carry, too.

If too much current should start flowing through the wire, the device is there to shut down the circuit before damage can occur.

The smaller the wire, the less current it can carry. Memorize the ampacity of typical wires.

#14—15 amps	#4—70 amps
#12—20 amps	#2—95 amps
#10—30 amps	#1/0—125 amps
#8—40 amps	#2/0—145 amps
#6—55 amps	#3/0—165 amps

These ampacities can be found in Table 310-16 of the Code book, along with many others.

FIGURING SWITCH RATINGS

Switches and the grounded-ungrounded switching situation are discussed in Article 380 of the Code book. The Code says, "...switching is done only in the ungrounded circuit conductor." It also says, "Switches or circuit breakers shall not disconnect the grounded conductor of a circuit."

However, Article 380-2 (b) presents two exceptions. These exceptions involve special applications as in a gasoline dispensing island, in which instance the neutral is opened along with ungrounded conductors. In ordinary circumstances, though, the neutral is not to be opened.

Switches connected in series with the ungrounded wires are rated in amperes. The same type of thinking that goes into calculating the rating of an overcurrent device is appropriate when calculating the rating of a switch.

For instance, a main switch must be able to carry the main current during the normal operation of the circuit. In a 200-amp panel, the main switch usually is rated at 200 amps.

An ordinary switch in a wall might be rated at 15 amps. The switch would be connected in series with a #14 wire. The overcurrent device for this circuit would also be rated at 15 amps.

Switches are also rated according to the voltage they can handle. For example, a switch can be designated 15A-277V. This means that the switch can be used in 120-volt circuits and in 240-volt circuits, but not in any circuit in which the voltage exceeds 277 volts.

The ratings are stamped on the switch, usually on the metal mounting strap. The Code also designates switches as being acceptable for AC-only use or for AC-DC general use.

The AC-only is the newer type. DC is not around as a common energy commodity any longer.

The AC-DC general use type is more expensive than the AC-only. Both types do the same safe, reliable job. Naturally, the AC-only type is the best choice. The DC feature on the more expensive switch is useless. Despite this, you'll find AC-DC general use switches in many buildings. You can replace them with AC-only when they go bad, as long as AC is being used there, and it almost always is.

RECEPTACLES AND PLUGS

Receptacles and plugs are also types of electrical devices. When you unplug a load from a receptacle, you actually are disconnecting the circuit. You're acting something like a switch or circuit breaker. As a matter of fact, on certain large appliances like ranges and clothes dryers, you are permitted to use the receptacle-plug system as a method of disconnection.

Meanwhile, notice we said that when you removed the plug from the receptacle that you were acting *something* like a switch. Actually, there is a major difference. A switch opens only the ungrounded line. A receptacle plug opens the ungrounded *and* grounded lines. As you pull the plug out of the wall, the neutral, as well as the hot line, disconnects.

How can that be? The Code specifically states that the grounded line must be intact. You cannot disconnect the grounded conductor with a switch or circuit breaker.

This is a different wiring situation. The switch and circuit breaker are in series with the wiring. The load, which is what is attached to the plug, is in parallel with the wiring. The load is an appendage to the wiring system. It is joined to the end of the wiring

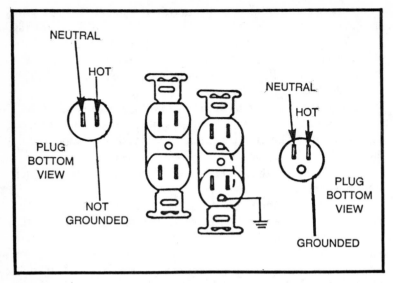

Fig. 8-6. The ground is attached to a special contact on the plug. This automatically grounds the appliance.

run. This topic will be discussed further in Chapter 11 when we examine series and parallel circuits.

Plugs and receptacles are rated according to their voltage and amperage capacities. Most of the common residential receptacles and plugs are rated at 600 volts or less and at between 10 and a few hundred amperes. They can satisfy all ordinary demands.

GROUNDING AND NON-GROUNDING TYPES

The Code requires that all replacement 15- and 20-amp receptacles be of the grounding type. The ground wire is attached to a special contact on the plug, as the illustration shows (Fig. 8-6). As you plug in an appliance, it is automatically grounded.

The grounding takes place in the third hole. This act of grounding is in addition to the grounding of the neutral wire. However, the neutral wire is part of the circuit and the grounding in the third hole is not.

The wire (prong) that comes from the appliance and inserts into the third hole is connected to the frame of the appliance, not to the appliance load circuit. When you plug into the third hole, you are grounding the appliance frame. Because of this, the frame remains at zero volts.

If a fault should occur between the appliance circuit and the frame, you will not get shocked if you touch the frame. Third-hole grounding is covered in greater detail in Chapter 10 (Grounding).

Over the past years millions upon millions of non-grounding receptacles have been installed in buildings. The receptacles are still there. They have only two holes, one for the hot line and one for the neutral. They offer no hole for the grounding of an appliance frame.

You cannot directly install the three-prong plug of your portable drill into one of the old-fashioned, non-grounding receptacles. You need an adapter that will translate a three-element arrangement into a two-element one. Yes, if a fault develops in the drill, and the frame becomes live, without the three-hole grounding system you'll get a shock. Yet you've been taking the chance that no fault would occur and you've used your drill, anyway. We've all done it.

Since so many millions of the old receptacles are still around, the NEC has felt compelled to permit the manufacturer of non-grounding receptacles to continue. Nevertheless, at every opportunity you should replace the non-grounding receptacles with the grounding ones.

Merely substituting a three-hole receptacle for a two-hole one is meaningless unless a circuit-grounding system is in effect. Without the circuit-grounding system, someone could come along and plug in an appliance with a false sense of security. Alas! He could experience a painful surprise.

CONTROLLERS

The Code defines a controller as: "a device or group of devices which serves to govern in some predetermined manner the electric power delivered...."

The definition is so general as to be almost worthless. You must examine the different kinds of controllers in order to understand them.

You are all famliar with a dimmer light switch. When you install one in the wall instead of an ordinary toggle switch, you are exchanging a switching device for a controlling device.

Not only is the dimmer switch able to turn the light off and on, it is also able to control the amount of wattage that flows into the light. With such a control you can set the light at bright, dim, or at any intensity in between.

A transformer in an electric train set can be switched on and off. However, it has various steps in between the extremities that can be

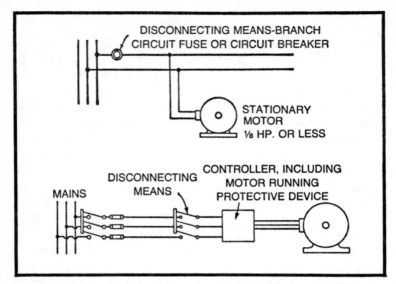

Fig. 8-7. Different methods of disconnection for motor circuits.

used to vary the power. The train can be speeded up by feeding it more power or slowed down by reducing the power it receives.

The dimmer switch in large public buildings such as movie houses or theatres can be a large rheostat. A rheostat is a wire-wound or carbon-resistive element with a movable contact arm. As you rotate the contact arm, you vary the resistance in the circuit. When you diminish the resistance, you allow more current to pass and brighten the lights.

As you put more resistance into the circuit, more of the power that was intended for the lights is absorbed. The lights grow dimmer.

In recent years, silicon controlled rectifiers (electronic devices) have replaced rheostats to a large degree. They are more efficient than the rheostats.

An autotransformer is used as the active element in the electric train set. It is a coil of wire with contact points and a wiping arm that jumps from contact to contact. This varies the resistance and inductance in the circuit and raises and lowers the power output.

MOTOR CONTROLLERS

A motor circuit requires a controller. The controller's job is to start and stop the motor. By performing these functions, the controller qualifies as a switch or circuit breaker. However, it has no ability to vary the wattage output. It is used on refrigerators, water pumps and other equipment to start and stop them automatically.

The automatic feature can be a switch that turns on or off when activated by a thermostat or water flow. Motor controllers are covered in Article 430-81 of the Code.

Motors of ⅛ horsepower or less, such as an electric clock motor, need no controller. The normal overcurrent protection in the circuit is sufficient. Motors of 1/3 horsepower or less can use the plug and receptacle as the controller. Motors of 2 horsepower or less can use an AC-only toggle switch, if the switch's amperage rating is at least 125% of the amperage rating of the motor. (The motor ratings and calibrations are covered in more detail in Chapters 15, 19 and 20.)

Figure 8-7 compares the methods of disconnecting a motor circuit: branch circuit fuse or circuit breaker versus controller device.

LAMPHOLDERS

The lampholder device is more commonly known as a socket (Fig. 8-8). Every lampholder has to have a socket, but strictly speaking the socket is only part of the lampholder. However, it is the part that makes contact and has the current flow directly through it. The socket is what makes a lampholder a device.

The rest of the lampholder is an assembly consisting of protective and insulative coverings (Fig. 8-9). The lampholder also includes

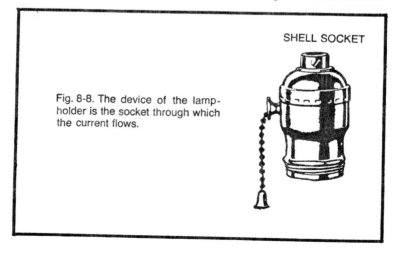

SHELL SOCKET

Fig. 8-8. The device of the lampholder is the socket through which the current flows.

175

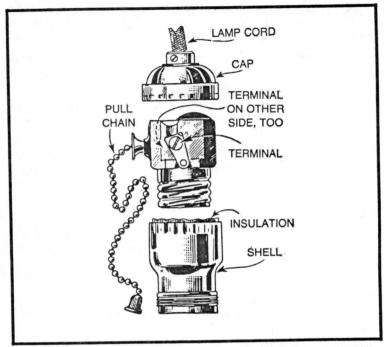

Fig. 8-9. The rest of the lampholder is an assembly which contains and insulates the socket.

the supporting hardware by which the lampholder is securely attached to the source of the current.

The terminals of the lampholder are called the base. There are a number of types of bases. Three common types are:

1—Shell
2—Mogul
3—Candelabra

If you look around your home, chances are you'll see all three types. Most conventional lamps have a shell base. The mogul base looks like the shell base but is appreciably larger. The candelabra base is tiny compared to the mogul and shell bases.

A typical brass shell lampholder consists of a shell and a cap. Sometimes the cap is called the base. The shell and cap snap together by means of the fluting which is molded into the lip of each. The fluting makes the socket easy to separate and snap together. The fluting completely encircles each lip and is able to hold the body and cap together when the chain is pulled with normal force. When more than average pressure is involved, however, the socket is apt to separate.

The manufacturers in the electrical field have cooperated in a policy of standardization. As a result, caps and shells are of a closely similar configuration and almost any cap can be used with any shell so long as they are of the same general type.

Bodies are made with brass enclosures and sockets on a porcelain or composition support. A fiber casing completely insulates the live socket from the brass enclosure.

Bodies can have any of several types of built-in switches. The switches can be operated by pushbutton, pullchain or tumble key. Some bodies have no switch at all; they are called "keyless."

Caps have threaded holes. The threads are conventional and you can get caps to fit pipes with diameters of ⅛", ¼", ⅜", and ½". Caps are also constructed with porcelain or bushing types of holes.

Bases are also manufactured to be mounted directly onto outlet boxes and for attachment to exposed wiring. All in all, the manufacturers of electrical fixtures offer a broad range of lampholders and accessories for many different kinds of installations.

DETAILS ON RECEPTACLES

As discussed earlier in the chapter, receptacles are plug-in convenience devices. The receptacle in the wall and the plug on the appliance or lamp make these electrical furnishings fully portable. Besides, the receptacle and plug arrangement provides an excellent way to disconnect appliances, especially large appliances such as ranges and clothes dryers. They attach and detach quickly.

The basic receptacle and plug (Fig. 8-10) are a pair of metal contacts that mate together securely. They "wipe" each other as they are joined. During their union considerable surface area is in contact.

The plug plays the male role in the union. Both rubber and plastic styles of plugs are manufactured. Some models of plugs have cord clamps whose function is to transfer from the conductors to the insulation the stress experienced when the cord is tugged to remove the plug from the receptacle.

The receptacle is the female component of the plug/receptacle union. Two models of receptacles are available: single-wipe contact and double-wipe contacts.

In the single-wipe model, one side of the plug reception cavity serves as a conductor and the other side as an insulator. The plug blade wedges tightly between the two surfaces.

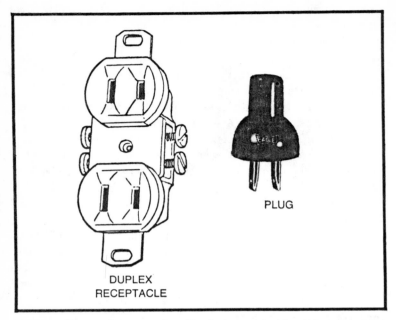

PLUG

DUPLEX
RECEPTACLE

Fig. 8-10. Receptacles are produced in single- and double-width models. Most 15- and 20-amp receptacles are of the duplex type.

Both sides of the plug reception cavity on the double-wipe model act as conductors. This doubles the surface area available for the passage of current from plug to receptacle. The double-wipe model is more expensive than the single-wipe model but is more durable because it is resistant to the carbonization which occurs at the moments of contact and disconnection.

Binding screws or pressure locking terminals or both are used to attach wire to 15- and 20-amp receptacles. When a conductor is to be attached to a binding screw, the end of the wire should be stripped of insulation and then bent around the shank of the screw in a clockwise direction. As the screw is turned down, the wire will tighten securely around the shank.

Pressure locking terminals are recessed inside the receptacles. When the stripped wire is thrust home into the plug reception cavity, spring clamp contacts will clasp it firmly.

DUPLEX RECEPTACLES

Duplex receptacles far outnumber single receptacles. Most of the residential duplex receptacles are of the 15- and 20-amp varieties. The modern era duplex receptacle is a very satisfactory device.

The blades of the receptacle are of the parallel type and the 125-volt size. As you examine the receptacle, you'll see two white terminal screws on one side and two brass terminal screws on the other side. If it is the grounding type of receptacle, a green grounding terminal will appear on the end.

The two white terminals are shorted together with a piece of metal. The two brass terminals are shorted together, too. Because each set of terminals is shorted together, you can enliven both ends of the duplex receptacle by attaching a single wire to either of the white terminals and a single black wire to either of the brass terminals.

There are a couple of other wiring arrangements you can make also. For example, you can remove the two shorting jumpers. Each end of the duplex receptacle is then isolated from the other. If you wish, you can then wire each end of the receptacle into its own circuit.

To use the receptacle in another way, first break the jumper connecting the brass terminals. This isolates the brass terminals from one another but the white terminals remain shorted together. Then you connect one white wire, which is neutral, to the shorted-together white terminals. Next you connect two black wires to the two separated brass terminals.

Why do you do this? You do this so you can put a switch in series with one of the black wires. Thus you now have a duplex receptacle with one switched end and one constantly live end.

HEAVY DUTY RECEPTACLES

Industrial plants have receptacles with ratings as great as 400 amps at 600 volts. In the home you'll find receptacles rated at as much as 50 amps at 250 volts. These receptacles are used for ranges, clothes dryers, air conditioners, etc.

Figure 8-11 illustrates a wide range of heavy duty receptacles now available. Notice the variety of the configurations. Each model is designed to accept no other plug than the appropriate one.

The electric range plug/receptacle combination has its own unique configuration. The third cavity is for grounding the range's frame.

NEMA CONFIGURATIONS FOR LOCKING TYPE PLUGS AND RECEPTACLES

Refer to individual listings in catalog for complete descriptions of catalog numbers listed below.

			15 AMPERE		20 AMPERE		30 AMPERE	
			RECEPTACLE	PLUG	RECEPTACLE	PLUG	RECEPTACLE	PLUG
2-POLE 2-WIRE	125 V	L1	Eagle # 135 802 870 806	Eagle # 2472 3472				
	250V	L2			Eagle # 846 806 894	Eagle # 824 874		
	277V A.C.	L3						
	600V	L4						
2-POLE 3-WIRE GROUNDING	125V	L5	Eagle # 2483 2493 3493	Eagle # 2473 3473	Eagle # 2343 2353 3353	Eagle # 2363 3363		
	250V	L6	Eagle # 2484 2494 3494	Eagle # 2474 3474	Eagle # 2344 2354 3354	Eagle # 2364 3364		
	277V A.C.	L7	Eagle # 2485 2495 3495	Eagle # 2475 3475	Eagle # 3355	Eagle #		
	480 V	L8						
	600V	L9						
3-POLE 3-WIRE	125/250V	L10			Eagle # 2345 2355 3355	Eagle # 2365 3365		
	3Ø 250V	L11			Eagle # 2346 2356 3356	Eagle # 2366 3366		
	3Ø 480V	L12						
	3Ø 600V	L13						
3-POLE 4-WIRE GROUNDING	125/250V	L14			Eagle # 2445 2455 3455	Eagle # 3465 3465		
	3Ø 250V	L15			Eagle # 2446 2456 3456	Eagle # 2466 3466		
	3Ø 480V	L16						
	3Ø 600V	L17						
4-POLE 4-WIRE	3Ø Y 120/208V	L18						
	3Ø Y 277/480V	L19						
	3Ø Y 347/600V	L20						
4-POLE 5-WIRE GROUNDING	3Ø Y 120/208V	L21						
	3Ø Y 277/480V	L22						
	3Ø Y 347/600V	L23						

Fig. 8-11. Heavy duty receptacles come in many sizes and shapes. They are keyed to accept only a particular type of plug.

DETAILS ON SWITCHES

It was mentioned earlier in this chapter that switches (Fig. 8-12) are rated in amps and volts. The Code requires you to match the switch with the wire and overcurrent device in each circuit. For instance, if a circuit with #14 wire uses a 15-amp circuit breaker, then you must use at least a 15-amp switch. The voltage in this type of circuit is usually 120 volts and the switch is usually rated at 125 volts at least.

While this type of installation is satisfactory, if at all possible you should provide the switch with a margin of safety. Use a 20-amp switch to carry the 15-amp circuit load.

The above procedure is applicable to most lighting and appliance loads. However, there are exceptions. If the load is a motor, transformer, fluorescent lamp or other load that contains windings of wire on a steel core, you'll need greater capacity. (This does not include small motors in small appliances.)

When wire is wound on a steel core, the load becomes an inductive type. Later chapters will tell more about inductive loads.

When an inductive load starts up, it draws more than its rated running current. The switch must be able to endure that first surge of current. Therefore, for inductive loads switches are required which are rated at 125 percent of the rated running current. This means that a 20-amp inductive load would need a switch rated at 25 amps. 20 amps × 125% = 25 amps.

Don't be misled by high voltage ratings on switches. It is with amps that you are more concerned. If a switch is rated at 20 amps, it cannot carry more than 20 amps whether its voltage rating is 125 or 277. (Keep in mind, though, that a switch is not to be used in a circuit where the voltage will exceed the switch's voltage rating.)

POLES

A single-pole switch (Fig. 8-13a) has two terminals. As you flick the toggle from OFF to ON, the mechanism joins the two terminals together. When you flick the toggle in the opposite direction, the shorted circuit is separated and the current turns OFF.

Some common types of single-pole switches are the quiet, mercury operated, delayed action, touch, canopy, tumbler, pull-chain, and push button.

The three-way switch (Fig. 8-13b) is unlike the single-pole. You can recognize it by its three terminals. The three terminals are necessary to control lights from two locations, as from the top and

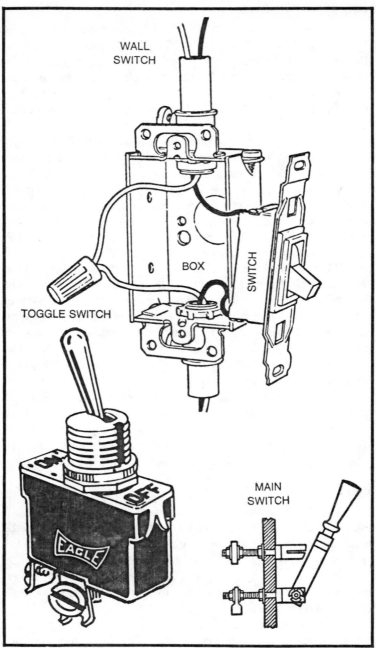

WALL
SWITCH

BOX

SWITCH

TOGGLE SWITCH

EAGLE

MAIN
SWITCH

Fig. 8-12. Switches come in all sizes and shapes. The wall switch, toggle and main switch are 3 common types.

bottom of a stairway, or at both ends of a long hallway. Three-way switches are normally used in pairs. There is very little use for just one of them. The actual wiring of a three-way switch is covered in Chapter 12 (Wiring Circuits).

The four-way switch (Fig. 8-13c) is somewhat more complex than the three-way switch. Its four terminals identify it. It is needed to control a light circuit from three or more locations.

Here is how to lay out a light circuit requiring switches at three or more locations. First of all, you need two three-way switches. One of the three-way switches is installed at the location nearest the light. The second three-way switch is installed at the location nearest the panel box and overcurrent device. All of the other locations in the circuit get four-way switches. If there are four locations, then you'll need two four-way switches. Five control locations require three four-way switches, and so on.

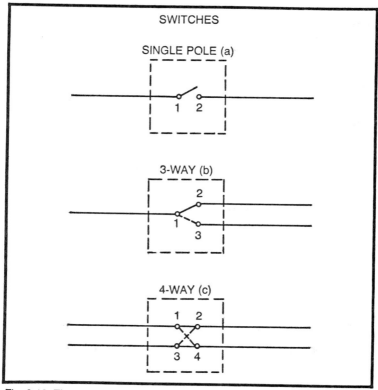

Fig. 8-13. The single-pole, 3-way, and 4-way switches are commonly used.

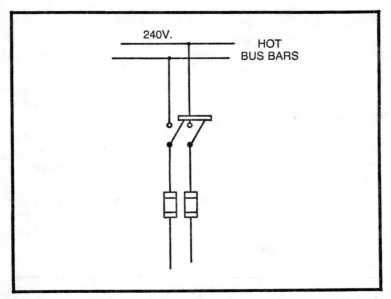

Fig. 8-14. The double-pole switch has four terminals. It is required on 240-volt circuits.

You probably noticed that we skipped the double-pole switch in the preceding discussion. That is because the double-pole switch is not used in routine house wiring. The Code requires its use in circuits that have no grounding. In non-grounding circuits you must open both conductors at the same time since they are ungrounded.

A double-pole switch has four terminals. Its handle opens (shorts) both sets of terminals. (Not both sets to each other.) Double-pole switches are required on 240-volt circuits (Fig. 8-14).

General Wiring Considerations

A complete wiring system is made up of components. For example, a home wiring job is composed of service entrance, branch circuitry, and ground. The branch circuitry, in turn, can be divided into individual circuits for the entrance, living room, dining room, bathrooms, kitchen, bedrooms and basement (Fig. 9-1).

THE WIRING SYSTEM

An electrical system can also be divided according to the hardware used. Some of the hardware components are conduit, conductors, fittings and devices. No component is any more important than the other.

The wiring system is a source of supply for electrical energy. The system supplies all the lighting and appliances in the home.

When you are capable of wiring a home, you are able to look at the proposed wiring installation and envision the correct components. Then your plans will come true as you install the components—in a manner to pass close inspections and meet all the safety requirements of the National Electrical Code.

Conduit has to be mounted in a sturdy fashion and filled properly. Conductors must be able to meet electrical safety standards on a permanent basis. Devices have to perform their switching and outlet functions, as fittings have to hold the entire system together securely.

185

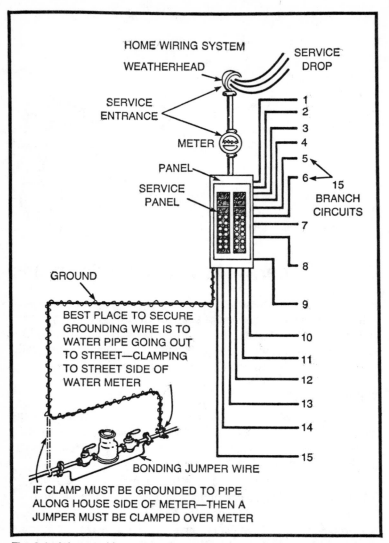

Fig. 9-1. A home wiring system is composed of a service entrance, branch circuits, and ground. The nature of the branch circuits differs according to their job.

The blueprint contains symbols showing the locations of components. Actually, much of the blueprint's data will not be used. You will change the job around somewhat as you work. You must know your component structure well, since in the final analysis it will be you who chooses the real materials and the final locations.

LOCATING DWELLING COMPONENTS

In the typical home wiring job the location of components is more or less spelled out in the Code. It's all been done before, and yet each job will have special aspects that you'll have to handle yourself.

In general, a new home has the following components:

1—Lighting outlets
2—Convenience outlets
3—Wall switches
4—Special purpose outlets
5—Individual equipment circuits
6—Kitchen electricity
7—Bathroom electricity
8—Outdoor electricity
9—Service entrance
10—Door signals
11—TV and radio antenna system
12—Telephone outlets
13—Correct grounding system

All of these components are described by the Code, the electrical industry, governmental agencies, and electric utility companies. Yet it is easy to go wrong, even though so much has been said about the components.

It is best to get the general rules down pat. Memorize the various positions of the equipment. It might be a good idea to make a check list. None of the rules is especially hard to understand. The difficulty is that there are so many of them. You'll have to be in command of these rules as you wire. You'll not be able to skip a step.

ABOUT THOSE LIGHTING OUTLETS

Good lighting is vital in a home. When you install the lights correctly, the home owner has optimum seeing standards. The eyesaving illumination is also decorative, esthetic and nice to live with.

The home has lighting fixtures, portable lamps and other lighting equipment. The outlets are to be installed so the illumination is comfortable, healthful and safe for all the work and recreational activities in the home.

Each room has its own kind of lighting. Some rooms need fixtures. Other rooms need a preponderance of portable lamps.

There are fixtures for the ceilings, walls, coves, valances, cornices, and stairways.

Fixtures are intalled to light up entrances, exits, hallways, porches, libraries, dens, dining rooms, etc. Each room requires thought.

The living room could include bookcase illumination and spotlights to brighten a painting on the wall. Bathrooms have medicine chests, nightlights and shower safety lighting. Bedrooms need fixtures for cosmetics tables, clothes closets and full-length mirrors. The kitchen sink needs light and the range must have some kind of fixture. It's also convenient to have lighting inside cabinets.

Hallways are safe when well lit. Basements need furnace lights, workbench lighting, and illumination for dark corners.

Each room has a character of its own. For instance, living room lights will not be appropriate in a bathroom.

CONVENIENCE OUTLETS

A convenience outlet is not a fixture outlet although it can be a lighting outlet also. All you have to do is plug a portable light into the outlet.

Convenience outlets are best located in corners of rooms. Furniture is more often than not placed midway along the walls. So that the outlets will be accessible, it may be a good idea to install them in the corners of rooms. Sometimes the corners do get closed up, but they're still a good choice.

Convenience outlets are usually located about 12 inches above the floor line. That measurement may be varied a bit according to circumstances, but it's been adjudged that 12 inches is a suitable height.

The Code says that receptacle outlets shall be placed in a room so that no point along the floor line of any wall is more than 6 feet from an outlet. The reason for this is that most lamp and appliance cords are not more than 6 feet in length. You should understand that when you avoid long extension cords, you eliminate safety hazards and wattage waste due to voltage drop in the extra length of conductor. The saving due to shorter cords won't be much, but every little bit helps.

While the three rules (installing outlets in corners, locating them 12 inches above the floor line, and placing them 12 feet apart with reference to the floor line) are to be followed as a guide, each room in a home can have exceptions or additions (Fig. 9-2).

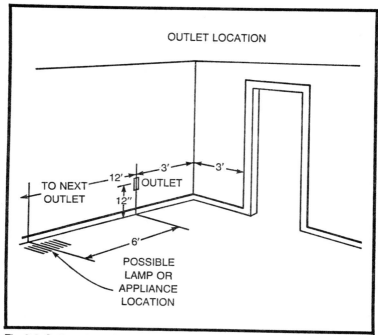

Fig. 9-2. Outlets should be installed near corners not more than 6 feet away from a possible lamp or appliance location.

You'll have to do some customizing to adjust the rules to the room in question. Kitchen and dining areas often present special problems. This situation is covered later in this chapter.

WALL SWITCHES

Practically all lights and a small proportion of convenience outlets are controlled by switches. Available are many different kinds of switches.

Toggle switches are placed on the wall. They are used to turn on lamps, either plug-in or permanent. The switch can be connected to the hot line of the fixture or to a receptacle outlet.

These switches usually are located at the latch side of doors. It is convenient and safe for the resident to reach into a room and light it up before he enters.

Even though most wall switches are installed within the room they are to illuminate, others are located outside a closed door.

An example of a switch located outside the door could be in the instance of basement illumination. By this arrangement one can light

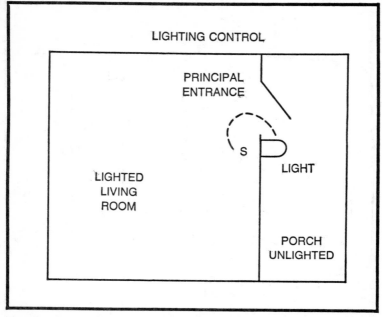

Fig. 9-3. Any entrance or exit should have an indoor switch to control light in an unlighted area.

up the basement before opening the door. This is handy if the door is at the top of a stairwell. An alternative method is to place the switch just inside the door, especially if there is a landing within.

Most outdoor lights have indoor switches. This protects the switch from the weather and allows the resident to turn on the lights from behind the safety of a locked door to inspect a stranger who has rapped for attention.

The wall switch is universally mounted 48 inches above the floor. This allows children as well as adults to operate the switch.

The 2-way, 3-way, and 4-way switches are mounted in agreement with the same general rules.

MULTIPLE SWITCHES

When two or more switches are needed to control a light, 3-way and 4-way switches have to be installed in the same circuit. The Code says that wherever wall switch controls are required and the entrances are more than 10 feet apart that a switch must be installed at each entrance. If the entrances are less than 10 feet apart, only one switch is required.

In actual practice, any entrance or exit at which a resident must move from a lighted into an unlighted area should have a wall switch to control the illumination of the unlighted area. An example would be the door leading from a living room to a porch. This is called a principal entrance (Fig. 9-3).

SPECIAL PURPOSE OUTLETS

The 15- and 20-amp duplex receptacles of the home are not special purpose outlets. All other types are. Special purpose receptacles are used for room air conditioners, portable space heaters, built-in ventilating fans, electric range, electric clock, food freezer, washer, dryer, furnace, cooling fan, etc.

Many of these loads are operated with 15- or 20-amp outlets, but the ordinary duplex receptacle is not used. Special purpose outlets are usually single outlets and get special wiring and possibly special ground arrangements.

As the name implies, each one is special and is designed to perform a certain job. They are considered heavy duty, even though the actual current drain may not be out of the ordinary.

INDIVIDUAL EQUIPMENT CIRCUITS

The branch circuits are divided into two general categories: Type 1 and Type 2 (Fig. 9-4).

Type 1—This is a branch circuit that is installed to supply but one load: a range, an air conditioner, a water heater, a dryer, a garbage disposer, a dishwasher, or a trash compactor.

The load is usually a fixed or stationary appliance. Article 100 of the Code defines appliances, fixed and appliances, stationary.

A fixed appliance is one which is fastened down and not intended to be moved. There might be plumbing attached to it, or it might be part of the building structure. Water heaters, garbage disposers, oil burner motors, room air conditioners, built-in ovens, cooktops, and so on are good examples. These appliances might even be installed with cord and cap for convenience, but they are fixed because of their plumbing or structural attachments.

A stationary appliance is one that is just sitting there and is not fastened to the plumbing, the building structure or anything else. The appliance *can* be moved, even if it takes a crane to do it. Self-contained ranges, window air conditioners, clothes dryers and latches are examples. They are connected with cord and cap and can

CIRCUIT TYPES

PANEL CIRCUIT #	TYPE		CIRCUIT TYPES	CODE REFERENCES
1	2	15		
2	2	15	FIVE CIRCUITS SHOULD BE PROVIDED FOR GENERAL PURPOSE	LIGHTING AND RECEPTACLES CODE 220-2
3	2	15	APPLIANCES SUCH AS EXTRA FANS, MOVIES, VACUUM, ETC,	
4	2	15	ON 15 AMPS	
5	2	15	FUSES	
6	2	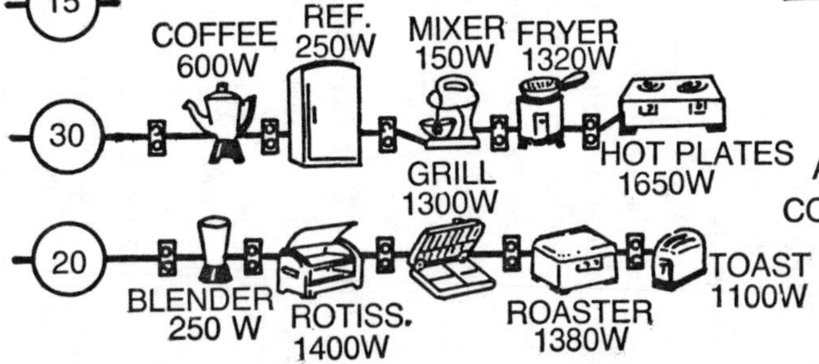 30	COFFEE 600W REF. 250W MIXER 150W FRYER 1320W HOT PLATES 1650W	2-20A SMALL APPLIANCE CODE 220-16(a)
7	2	20	BLENDER 250 W ROTISS. 1400W GRILL 1300W ROASTER 1380W TOAST 1100W	

192

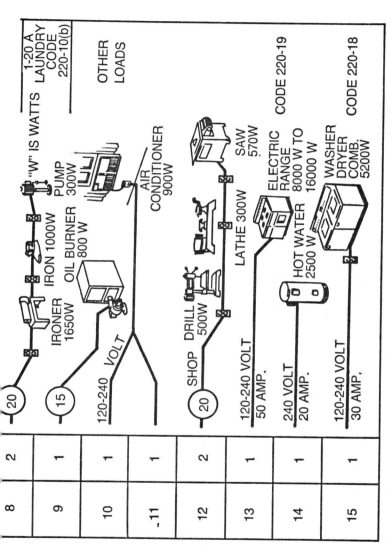

Fig. 9-4. There are two types of general circuits.

193

be removed from a dwelling without any problem, except for the effort required.

These appliances all could need Type 1 circuits (individual lines). The Type 2 circuits are designed to supply two or more outlets with electricity. These are the lighting and portable appliance load circuits.

KITCHEN ELECTRICITY

Chapter 14 (Residential Wiring) discusses house wiring room by room. The kitchen is one of the rooms discussed, and it is a different type of installation than the other rooms (Fig. 9-5).

The kitchen is different because it is the center of all the home electrical activity. Service panels are often installed near the kitchen since so much electrical equipment is located there and panels should be situated near the most active electrical systems.

Kitchens in recent years have become opulent and are lavishly equipped. The list of appliances ranges from a main intercom panel to an electric knife. Artificial illumination is abundant. If fluorescent fixtures are found anywhere in the dwelling, it'll be in the kitchen.

All of the general illumination is controlled by wall switches. The lighting is bright and the entire kitchen is illuminated thoughtfully. Counter tops, sinks, range, dishwasher, and so on need electrification. Nearby is the water heater and possibly the laundry.

Rules for convenience outlets prescribe at least one outlet for every 4 linear feet of counter top and at least one outlet for every individual work surface. The outlets for work surfaces are installed about 44 inches above the floor.

The refrigerator should have its own outlet, while not necessarily needing to be of a special type. Special purpose outlets are for the range, ventilating fan, dishwasher, garbage disposer, trash compactor, electric clock and, if it's in the kitchen, the food freezer. All the rest of the outlets are split receptacle types.

BATHROOM ELECTRICITY

The bathroom has also become unusual in design and application. It is no longer a tiny watercloset with an overhead light and a single receptacle. It is an institution. Appliances and lighting are everywhere in it.

Ceiling fixtures in line with the front edge of the wash basin give improved lighting at the mirror. When more than one mirror is in the room, additional lighting for each is needed.

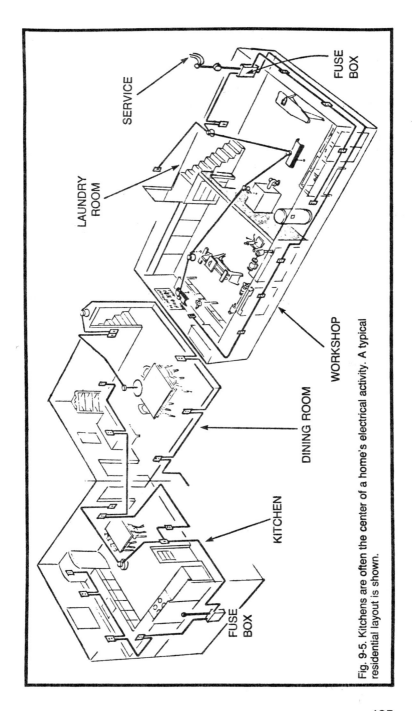

SERVICE

LAUNDRY ROOM

FUSE BOX

WORKSHOP

DINING ROOM

KITCHEN

FUSE BOX

Fig. 9-5. Kitchens are often the center of a home's electrical activity. A typical residential layout is shown.

The lighting at the mirror itself is carefully planned. Generous illumination of the human face and neck is the goal. Multiple light fixtures with unusual features are the rule.

Shower stalls have become more important than bathtubs, so they get vaporproof lighting and a wall switch just outside the stall.

Steam bath and whirlpool appliances are sometimes wired separately as garbage disposers and dishwashers are in the kitchen. Other kinds of appliances have to be plugged into convenience outlets. These outlets are not 12 inches above the floor line. They are usually installed alongside the wall switch or near each mirror at a point 48 inches above the floor.

A receptacle in a lighting fixture is useful, but it cannot take the place of a convenience outlet on the wall.

Special purpose outlets must be installed for the space heater and the ventilating fan. The fan outlet is switch-controlled either in combination with the ceiling light or by itself.

Switch-controlled night lights are recommended for the bathroom.

OUTDOOR ELECTRICITY

The definition of an outdoor area is confusing. There is no problem with the grounds, but is a screened-in porch outdoors or indoors?

The answer is hard to understand. It is sometimes considered indoors, yet the Code does not count it as living space when lighting needs are calculated. According to the Code a dwelling needs 3 watts per square foot. Yet the code says that computed floor area shall not include open porches. An open porch, therefore, is outdoors.

If the outdoor area such as a terrace or patio is not roofed, then the outdoor building wall should have some general illumination mounted on it. A post lamp is acceptable. A wall switch inside the house should control the light.

When the area is roofed and has more than 75 square feet of floor area, then it shall have a lighting fixture controlled by a wall switch. Should there be two or more doors leading out to the roofed area, each door would be well served with a wall switch to operate the light.

Floodlights are useful to illuminate the grounds. Colored and special effect lights add decoration besides providing illumination. These areas can be made beautiful with carefully placed lights.

Convenience outlets with weatherproofing are installed every 15 feet on walls that border porches, terraces, and patios. The outlets are most effective when they are controlled by switches mounted inside the house near the door.

SERVICE ENTRANCE

The "service" is the main component of an electrical system. It is the heart of the entire system. All of the lighting, outlets, switches, and other branch circuit constituents stem from the service.

The service is the entrance to the system and stands between the utility company's wires and the house wiring. The service has the weatherhead, wattmeter and panelboard in an overhead drop installation. An underground service has no weatherhead since the wires enter the house from below grade.

All of the current used by a house must pass through the service. The current is drawn into the weatherhead from the utility company's lines. The current travels down the service wires to the wattmeter. All of the current flows through the meter and is recorded. You pay the utility company for the number of kilowatt hours recorded on the meter.

The current then flows into the main switch. This is a switch that can disconnect the entire electric system when it is thrown. It is needed during repair time or emergency to stop the current at the source, before it can get into the house.

The current then goes to a distributing panel. In the panel the current is distributed to all the branch circuits. Each branch circuit has its own overcurrent device in the panel.

In homes, most service entrance installations are 3-wire, 120/240-volt arrangements. This type of service is usually adequate for homes with a floor area of up to 3000 square feet. You can choose the size of the service by the number of amperes required. It might be 100 amps for a tiny home, 125 amps for a good-sized apartment, 150 amps for a medium-sized home, and 200 amps for a large home.

Each size of service requires a different size of conductor. The larger the ampacity, the larger the conductor needed. The larger the wire, the larger the pipe needed to contain it. The larger the current, the larger the size of the main switch you'll need.

The service is installed at a convenient place in the house, preferably near the major appliances. The utility company will cooperate with you on locating the service. Their advice is usually excellent.

Sizing the service conductor is an important job to be done as the layout of the electrical system is planned. Besides the entrance conductor, a grounding conductor for the service must be sized.

DOOR SIGNALS

Doorbells, chimes, etc., are part of the wiring system. The bells and other sound makers are usually installed where they conveniently can be heard. The pushbutton switch, of course, is at the door. It should be near the door even though mounted on a post standing apart from the house.

The door signals do not use the full 120/240 volts. They operate on low voltage. This means 30 volts or less, possibly around 12 volts.

The low voltage is produced from the full voltage by a step-down transformer (Fig. 9-6). Up to the place where the transformer is attached to the full voltage, the Code must be met as usual.

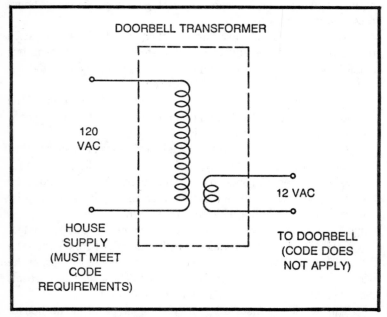

Fig. 9-6. The doorbell or chimes are powered with only 12 volts. A step-down transformer such as this one changes 120 volts to 12 volts.

However, once the 120 volts have been transformed to 12 volts, the Code does not care much about what is done.

The 12 volts present little or no electrical hazard. The important consideration is to keep the 12-volt wiring separated from the full-voltage wiring.

You are not permitted to run bellwire in the same pipe or armor as full-voltage wire. You are not permitted to place the low-voltage wire closer than 2 inches from the full-voltage wire, unless a porcelain tube sheathes the low-voltage wire. You cannot run the low-voltage wire into any box unless a metal barrier of the same thickness as the walls of the box is interposed between the full-voltage and low-voltage wires.

TV AND RADIO ANTENNAS

The Code (in Article 810) is specific about TV and radio antennas. However, the main consideration of the Code is safety and not antenna performance. The Code has no interest in whether the TV reception is good, bad, or indifferent.

First of all, the Code wants to make sure that the antenna lead-in wire and supports are nowhere near the electrical system. Secondly, the Code wants the supports to be secure and grounded properly.

Outdoors, the antenna shall not be attached to poles carrying power of more than 250 volts. When the voltage is less than 250, the same pole can be used, but antenna conductors and electrical conductors must be situated at least 2 feet apart.

When antenna conductors are strung from an antenna to a building, they shall not cross over electrical conductors. They shall be kept securely away so there is no possibility of them contacting each other.

The masts and metal structures which support the antennas must be grounded correctly. The best ground is the same water pipe to which the service ground attaches. A separate clamp for the antenna should be used. Don't use the clamp of the service entrance ground.

Indoors, the antenna or lead-in shall not be run nearer than 2 inches to other conductors. This includes the low-voltage wiring. It is not allowed near the antenna lead-in, either.

The reception problems are really of no concern to the electrician. That is the TV tech's province. Incidentally, electricians are sometimes called upon to run coaxial cable for antenna systems.

Coaxial cable is pulled exactly like any other cable. Just keep the cable away from the electrical system as the Code dictates.

TELEPHONE OUTLETS

Telephones are usually left entirely to the telephone company personnel. An electrician could probably get away with ignoring the telephone outlets altogether. Yet, to earn a name for himself, he might do these little extra things. This keeps him ahead of the competition and enhances his reputation.

While the walls are open, it is easy to run some ½-inch EMT and terminate it in switch boxes at the locations where the homeowner wants telephones.

The switch box can be covered with a wall plate that is fitted with a single bushed opening for the telephone cord.

Then when the telephone man arrives, he can pull the tiny wire pairs through the EMT and do a neat job. The electrician will get the credit and the cash for the extra bit of work.

Otherwise, the telephone man arrives after the walls are closed and he has to pull wires across rough structural surfaces and across the exposed run of electrical wire.

The Code book rules on telephone systems in Article 800 (Communications Circuits). In the home the only real consideration is to keep the telephone wires away from the electrical wiring.

If the telephone wires should accidentally contact full voltage, an extremely hazardous situation would arise.

GROUNDING

Grounding is one of the important parts of an electrician's mental repertory. He must learn it carefully so that he can ground his installation correctly.

Poor grounding is hazardous and can cause electric shock, fire, and explosion. Correct grounding makes the installation safe from those hazards.

Chapter 10 discusses grounding in detail.

OTHER COMPONENTS

In this chapter we have discussed the various general considerations that go into a residential electrical system. There are other components of the system that we haven't mentioned, components that are involved with special areas of the home like swimming pools, recreation rooms, fountains, and so on.

Residential isn't the only type of wiring, either. On the millions of farms and ranches, there are special kinds of electrical systems. Right outside your door there is street lighting. Other kinds of electrical jobs are being done not far away. In apartment houses, stores, shopping centers, and office buildings, there are commercial kinds of electrical wiring.

Industrial wiring is a complete electrical subject of its own. An electrician might be employed to operate in any of the above types of projects.

Most electricians work on home and apartment house systems. But no matter what kind of electrical system it is, the basic components are conduit, conductors, fittings, and devices—whatever the application.

At this point in the book, you should have a good idea of what electrical systems are made of. The next four chapters are going to examine the way that electrical systems work.

We'll start with grounding the system and then go into actual electrical circuits and what is happening in the circuits to cause the appliances and lights to operate.

By now you should appreciate just how valuable the National Electrical Code book is to everyone engaging in electrical work of any description. The professionals in the audience no doubt already have their own copies. Those of you who intend to embark upon electrical careers should purchase a copy at the earliest opportunity; after all, the Code book is going to be at your fingertips for the life of your career. And you do-it-yourselfers will find the Code book an eminently useful and practical guide in the conduct of your wiring projects. So, obtain your own copy of the latest NEC Code book and start right now to mark it up as you begin to understand its sections.

Grounding

The purpose of grounding is safety. When an electrical installation is not properly grounded, the danger is great. Improperly grounded electrical systems have been responsible for destroying all kinds of electrical equipment, for starting fires that consumed buildings, and for bringing death to men and animals by electrocution.

TWO DIFFERENT TYPES OF GROUNDING

In the parlance of the electrician's trade, there are two types of grounding. One is called *system* grounding and the other is called *equipment* grounding.

System grounding has to do with grounding the neutral wire of the electrical circuit. The neutral wire is usually the white insulated wire and it carries current just like the hot wire (Fig. 10-1).

Equipment grounding refers to the grounding of equipment frames, conduit, boxes, panels and the like. Either a green wire or a bare wire is used for this type of grounding and no current passes through exposed equipment or the green wire.

Depending on which type of grounding is employed, the white or the green wires will assume the same voltage potential as the earth. This potential is zero volts.

If a fault occurs in the electrical installation, such as occurs when insulation fails or hot wires accidentally contact the neutral

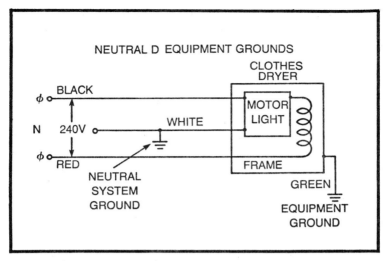

Fig. 10-1. The neutral system ground wire has current running through it during operation. The equipment ground normally passes no current.

conductor or equipment, the current is drained off to ground and cannot cause harm.

If the installation is not grounded, the neutral wire or equipment could become live. Anything or anyone touching the live installation would ground the current through themselves.

THE ACTUAL GROUND

When you study the Code, Article 250 (Grounding) appears quite complicated. However, residential grounding is not hard; it's easy to understand and employ.

The "ground" referred to in the Code is the grounding component. The ground is an electrode. It is buried in the earth. When the electrode is a rod, it is driven into the earth as shown in the illustration (Fig. 10-2).

When the ground is an underground pipe system and the buried portion is longer than 10 feet, the cold water pipe can do double duty and act as the grounding electrode while it carries water. The water pipe is composed of a conductive metal, runs long distances, and makes a permanent contact with the earth.

When the electrode is a rod, the neutral from the overhead wire into the building has a grounding wire which runs down the side of the building (or down the service pole) and connects to the rod.

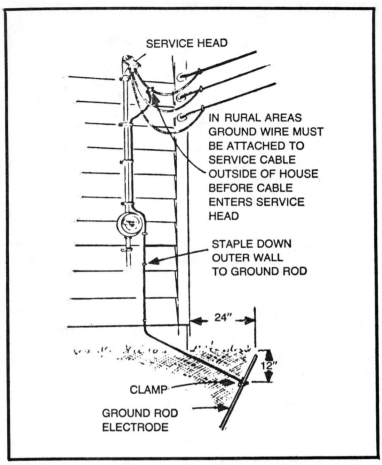

SERVICE HEAD

IN RURAL AREAS
GROUND WIRE MUST
BE ATTACHED TO
SERVICE CABLE
OUTSIDE OF HOUSE
BEFORE CABLE
ENTERS SERVICE
HEAD

STAPLE DOWN
OUTER WALL
TO GROUND ROD

24"

12"

CLAMP

GROUND ROD
ELECTRODE

Fig. 10-2. The common means to ground an electrical system is with rods or the water pipe system.

If the electrode is the cold water pipe, then the neutral is attached to the neutral bus bar in the panel and a grounding wire is run from the panel to the pipe.

A cold water pipe with a buried portion shorter than 10 feet is not a correct electrode, and another approved grounding electrode must be added to it.

THE WHITE WIRE (NEUTRAL)

When the ground is an underground pipe system and the buried portion is longer than 10 feet, the cold water pipe can do double duty and act as the grounding electrode while it carries water. The water pipe is composed of a conductive metal, runs long distances, and makes a permanent contact with the earth.

When the electrode is a rod, the neutral from the overhead wire into the building has a grounding wire which runs down the side of the building (or down the service pole) and connects to the rod.

If the electrode is the cold water pipe, then the neutral is attached to the neutral bus bar in the panel and a grounding wire is run from the panel to the pipe.

A cold water pipe with a buried portion shorter than 10 feet is not a correct electrode, and another approved grounding electrode must be added to it.

THE WHITE WIRE (NEUTRAL)

The white wire has been mentioned over and over again. The white, of course, refers to the color of the insulation enveloping the metal conductor. The white wire is the companion to the black and red wires in a home circuit. The black and red are each about 120 volts above or below (away from) ground. Ground is zero volts or neutral.

Black is 120 volts on one side of zero and red is 120 volts on the other side of zero. Between black and red, there is a difference of 240 volts. Each side, though, differs no more than 120 volts from ground.

The Code requires that the neutral wire be clearly apparent and be designated by a white or light gray color. It is required practice (with only a few exceptions) that white wire never be used for any wiring purpose other than as a neutral. The white (or gray) neutral wire is the only one authorized by the Code to serve as the system ground.

In actual practice, however, you'll sometimes substitute white wire for black and black wire for white. When you do, be sure to paint or tape the ends of the black wire white and the ends of the white wire black. As long as you clearly identify the wires, the circuit meets Code requirements.

The white neutral wire must NOT be interrupted at any point along its length. No switches, circuit breakers or other devices are permitted to break the continuity of the system ground.

Many 240-volt circuits do not use a neutral. That's okay, because they don't violate the taboo of opening up the neutral.

When you ground the neutral, you are joining together the plumbing system and the electrical system at zero volts. Since you, standing on earth, are also at zero volts, there is no voltage potential between you and the metallic piping of the home.

GREEN GROUNDING WIRES

"Green" is the name for the wires with the green insulation. An electrician may familiarly refer to it as #12 green. The actual insulation may be either entirely green or green with yellow stripes (the European color code for green).

These green wires are the grounding wires attached to appliance frames and receptacle ground terminals. In an appliance cord, the green wire is the third wire. In a receptacle, it is attached to the third hole.

Grounding wires can also be bare with no insulation at all—just plain copper or aluminum conductors. The absence of insulation poses no danger since the wires are grounded and there is no current in them. The bareness of the wires offers more surface area for contact against grounded pipe and boxes.

You are not permitted to use green to carry current or to serve as a neutral wire. Green is made to be used only as equipment ground wire.

Years ago, appliances were made with only two wires: a black and a white (the hot and the neutral). This is a potentially dangerous situation. Should the hot wire develop a short to the frame of an appliance, the frame becomes live and dangerous.

For example, Fig. 10-1 shows an appliance with a heating element. The element has 240 volts across it. As long as the element remains away from the metal frame, there is no danger. You can safely touch the frame because it contains no voltage.

Suppose, though, a short develops between the element and the frame. The short does not get drained off to zero volts. The frame simply gets hot with 240 volts on it. The element keeps operating (Fig. 10-3a).

You touch the frame and blam! You get knocked across the room, or worse.

If the cord has a green third wire attached to the frame, however, the danger is eliminated. The frame stays at zero volts. A lot of current could flow through the short and blow the overcurrent device, but that is the worst that could happen (Fig. 10-3b).

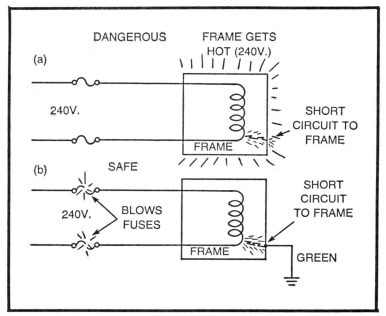

Fig. 10-3. (a) Without an equipment ground, an appliance is a hazard. (b) With an equipment ground, an appliance is safer.

RECEPTACLE GROUNDS

The third hole in the 120-volt receptacles is the grounding terminal. All the grounding terminals are wired together with green. The other two terminals are for black and white.

When conduit or armor cable is used, the terminal is connected to the box and the metal takes the place of green.

If cable is used without armor, it can contain an extra bare wire. The bare wire is acting as the green wire. Whatever the system, the third terminal is grounded. Therefore, when a three-prong cord from an appliance is plugged in, the third wire from the appliance frame is automatically and safely grounded.

The Code requires three-prong cords on every refrigerator, freezer, air conditioner, clothes washer, dishwasher, dryer and sump pump. All hand-held, motor-driven tools (such as drills, sanders and saws) are required to use three-prong cords. The Code does not require three-prong cords on TV sets, razors, vacuum cleaners, lamps, and so forth.

When the frames of these appliances are plastic, there is very little hazard. On the other hand, an interior electrical fault could turn

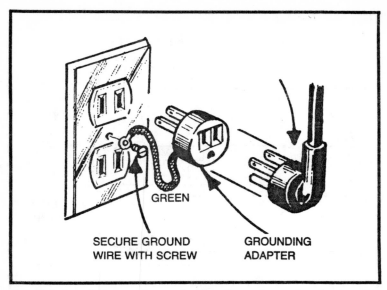

Fig. 10-4. An adapter can be used with a three-wire cord and a two-wire receptacle.

a metal frame live. Three-wire systems should be used if at all possible.

Many long-installed receptacles are not grounded. Most of them only have two holes. To connect a three-prong appliance like an electric hand drill to a two-hole receptacle, you can use an adapter (Fig. 10-4).

The three-to-two adapter shown in the illustration has a short green wire with a small brass ring on the end. The ring is fastened with a screw to the receptacle's flush plate. This arrangement can provide a direct connection to ground through the receptacle and box.

When you use a three-to-two adapter, however, keep in mind that you might be canceling the operation of the appliance cord's ground wire. Chances are that nothing will happen. But if a short should develop in the appliance, you could get shocked when you touch the case.

CONTINUITY

To an electrician the concept of grounding embodies the grounding of *all* of the exposed metal in a building—the only exception being the electrical conductors themselves. This includes pipes, cabinets, equipment, and even the window screens. When all of the exposed metal is grounded, everything is at zero volts. Without voltage potential, current cannot cause damage.

Residential electrical systems are grounded-neutral. There are other systems, of the industrial type, that are ungrounded.

Grounded-neutral means that all conduit, armor and any exposed metal in an electrical system are grounded. As mentioned before, the equipment ground method utilizes a green wire and the neutral ground method a white (or gray) wire. The green and white wires are strung alongside each other and meet at the house panel. When an electrical system uses conduit, the tubing and boxes can be substituted for the green wire. Whatever the composition of an equipment ground, the path leads to the panel.

As long as the equipment and system grounds are continuous throughout the installation, the maximum amount of safety is derived. If a conduit should accidentally become detached from the ground, a hazard is born.

Take the instance in which a piece of conduit at the end of a laundry room run becomes detached from the rest of the system. Until its separation, the conduit had been joined to an outlet box containing a receptacle.

Meanwhile, the insulation around a 120-volt hot line in the laundry room has deteriorated. Somehow, the ungrounded conductor makes contact with the conduit (Fig. 10-5).

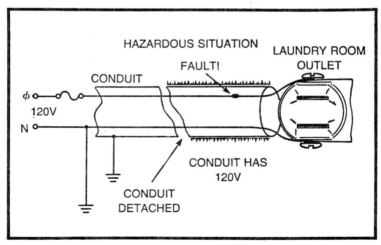

Fig. 10-5. If the ground was intact throughout the conduit, the fault would open a fuse and prevent the pipe from getting hot.

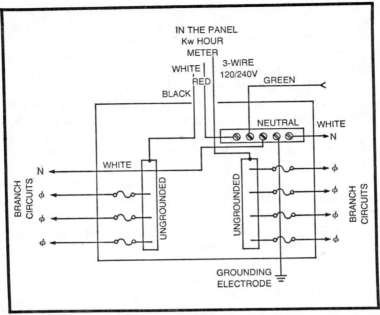

Fig. 10-6. The panel is grounded to the water pipe. The equipment grounds and system grounds all meet here.

If the ground had been intact, the short circuit would divert a lot of current from the conductor to the conduit. The fuse to the laundry room would blow. (All replacement fuses would also blow—until the fault has been remedied.)

Since the ground is not intact in our instance, however, no current flows. The conduit just becomes live. The first person to touch the conduit will return the current to ground through himself. If he is standing on a wet surface (a likelihood in a laundry room), he will sustain a shock which could even be fatal.

IN THE PANEL

Attached to the wall of the panel is a neutral bus bar. It is usually made of copper. It is fitted with a number of terminals.

When the common three-wire, 120/240-volt service is delivered, the three wires enter the panel through a knockout from the watt-hour meter. Typically, the wires are black, white and red. The white is neutral and the others are ungrounded conductors. The white is run directly to the bus bar and screwed down tightly. The white is at ground—zero volts.

The ungrounded conductors attach to the two ungrounded bus bars that are situated in the panel near the neutral bus bar. The ungrounded bus bars, in turn, are joined to overcurrent devices.

The branch circuit conductors also enter the panel through knockouts. The black and red wires of the branch circuits also attach to the overcurrent devices.

The white wires of the branch circuits pass directly to the neutral bus bar (Fig. 10-6). The white wires don't get switched or fused. They attach at the same point as the service wire. There is no break in continuity.

Through a different knockout passes another wire. It is the grounding conductor. It attaches to the neutral bus bar, too. The opposite end of this wire is joined to the water pipe or to some other grounding electrode.

Table 250-94 of the Code presents data on the sizing of grounding conductors.

BONDING CONDUIT

In any installation in which it is used, conduit is bonded to the panel (Fig. 10-7). Bonding is accomplished with copper wire and a grounding bushing.

The bushings connect the conduit to the panel. Each bushing has a setscrew. After the screw has been set, a copper jumper wire is attached between the bushing and the neutral bus bar. The bushing and locknut constitute a good ground connection in them-

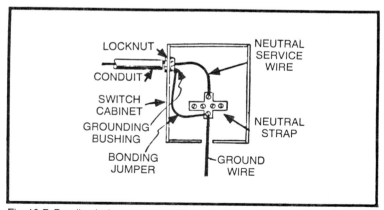

Fig. 10-7. Bonding is the permanent joining together of the metal components of an electrical system to insure continuity. A bonding jumper is a wire that connects various sections of the system together. The jumper electrically shorts out any possible resistances such as a water meter or a piece of flexible conduit.

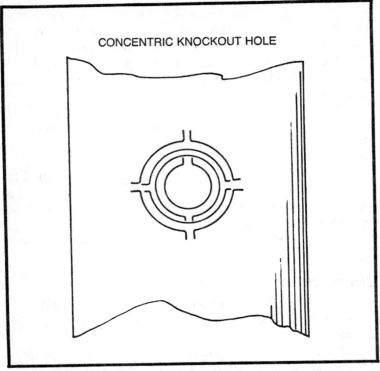

Fig. 10-8. Remove the center ring to obtain a small aperture or one of the outer rings for a larger aperture.

selves. The bonding insures a safe system even if the thread contacts corrode.

The connection of the conduit from panel to boxes is as important electrically as it is mechanically. When rigid pipe is used, care must be taken to secure it strongly at the junction points. When EMT is used, the compression, setscrew, or in-dent must be drawn up tight.

When conduit is fastened to boxes, the locknuts should be drawn up until they bite into the metal.

KNOCKOUTS

There is more to a knockout than there seems to be at first glance. A knockout requires a bit of technique. One style of knockout consists of a simple disc, but another style consists of a series of concentric rings (Fig. 10-8).

Each ring is a knockout of a particular diameter. The usual diameters are ½″, ¾″, 1″ and 1¼″. The electrician is interested only in the ring with a diameter the same as his conduit.

Since the efficiency of a concentric-style knockout as a ground connection is somewhat dubious, special care should be taken to establish a top-notch mechanical bond.

The rings should be removed one at a time until you arrive at an opening of the size you desire. Use a screwdriver and pry each ring toward you. With your lineman's cutters clasp the ring and bend it back and forth. When metal fatigue occurs, break off and remove the ring.

If you err and remove too many rings, don't discard the box. You can use reducing washers to restore the proper aperture.

GROUNDING CABLE

When nonmetallic sheathed cable (instead of conduit) is used, you must make your equipment ground of green or bare wire.

The green wire is attached to the neutral bus bar in the panel and is strung alongside the cable throughout the installation.

The green wire enters every box. It is attached to the grounding screw on the receptacle. Then a second green wire is extended from the first to a grounding clip on the box.

Still another green wire is attached to the original green and is run to the next box. With this type of installation a receptacle can be removed without interrupting the grounding of receptacles beyond. You are cautioned never to cause a disconnect in a grounding system.

THE GROUND FAULT INTERRUPTER (GFI)

When is voltage to ground? The Code says voltage to ground in an "ungrounded circuit is the maximum voltage between any two wires in a circuit" (Fig. 10-9a).

Home electric systems are grounded. Therefore if you have voltage between the hot line and neutral, you have voltage to ground.

What causes a fault? A fault in an electrical system is commonly caused by the contact of wires whose insulation has deteriorated. If a bare hot line touches a bare neutral line, a ground fault occurs. Current starts flowing abundantly and the energy is squandered. Not only that, the situation is dangerous.

Electrical systems contain fuses and circuit breakers. When a fault occurs and too much current (50 amps in a 20-amp circuit) starts to flow, the fuse blows (Fig. 10-9b).

You know this, but suppose you develop a fault that draws only 10 amps in a 20-amp circuit? The fuse holds. The flow does not stop and it is potentially lethal (Fig. 10-9c).

If you are standing outdoors in the rain when you contact the 10 amps, you will probably be electrocuted. What can be done?

You can install in the circuit a special device. It is called a ground fault interrupter (GFI). The GFI will break the circuit when the circuit's fuse, because of its larger tolerance, won't.

LEAKAGE CURRENT

A leak in a circuit is a fault, but not a short-circuit type of fault. While a short circuit will activate the overcurrent device, the leak might not. The virtue of the GFI is that it can sense the leak and break the circuit before a lethal amount of current can be loosed.

GFI's are relatively new devices but have rapidly achieved respect. Article 210-8 of the Code book says that GFI's shall be installed outdoors and in bathrooms. These are the areas where leakage current can be most dangerous. In the bathroom you are likely to be barefooted and standing on moist or wet surfaces. The outdoors is subject to rain and snow.

All appliances have a certain amount of leakage current. Usually it is too small to be measured. Some of the current will pass through the insulation and flow into the frames and cases of appliances and fixtures. This current constitutes a fault and can create an unwanted circuit through your body (Fig. 10-10).

The leakage current does not pass through the neutral. In normal circumstances the hot line and neutral line of a small appliance have the same amount of current passing through them.

In a leakage situation the hot line continues to pass the same amount of current, but the neutral has to share its flow with the leakage path.

HOW THE GFI SENSES

The GFI constantly monitors the amount of current flowing through the hot and neutral lines. The GFI assumes the place of an ordinary receptacle. The appliance is plugged into the GFI (Fig. 10-11).

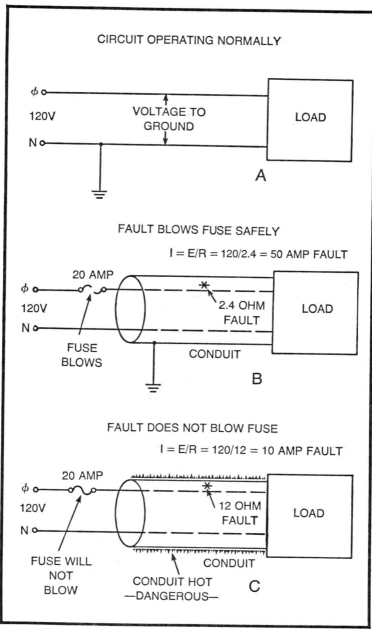

Fig. 10-9. The 20-amp fuse will blow if the fault causes more than 20 amps to flow from the hot line to the conduit. It will not blow if the fault draws less than 20 amps; this is a hazard.

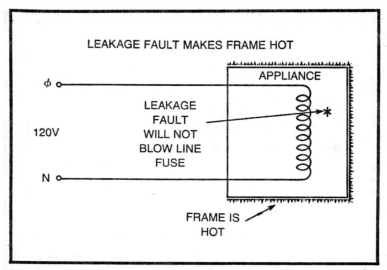

Fig. 10-10. In all appliances there is a small amount of leakage. If the leakage is large enough, the frame will become hot.

The GFI is able to sense when an unbalance between the hot line and the neutral occurs. It does this by way of a specially built toroidal coil. The coil is wound around both the hot and neutral wires.

As the current flows along the normal circuit, a magnetic field is developed around the wires. Since the current is flowing one way in the hot line and the other way in the neutral at the same time, the two magnetic fields are equal but opposite. The fields cancel out any current that would tend to be induced in the coil. (Actually the fields are just a bit unbalanced but the GFI is adjusted to compensate for that.)

When a ground fault occurs between the hot wire and the appliance or fixture, the neutral current flow drops off. This reduces the field around the neutral wire. The unbalance between the larger current in the hot line and the smaller current in the neutral induces a current flow in the sensing coil.

The coil thus detects the ground fault and sends the unbalance signal into a differential transformer and a solid state electronic circuit. This action enlivens another coil and a special circuit breaker snaps open.

The GFI is designed for LIFE PROTECTION. It is able to trip the breaker and open the circuit in less than 25/1000 second. Even though the victim would sustain a shock, the duration of the experience would be so short that he would not be seriously harmed.

Certain types of GFI's are designed for EQUIPMENT PROTECTION. These are industrial types.

THE ACTUAL GFI

Ground fault interrupters come in a number of trade shapes and sizes. Those used in residential installations are made as substitutes for receptacles—single or duplex. The GFI can be installed in a box, or simply plugged into an ordinary receptacle. When it is plugged into a conventional receptacle, the appliance is plugged into the GFI.

GFI's are rated like any device. Typical GFI's are 15- and 20-amp types like ordinary receptacles. The rating denotes how much current can safely flow in the device.

Various types of GFI's are now on the market. There is the weatherproof type which is designed as an outdoor receptacle. It is meant for permanent installation and looks like a duplex receptacle. Don't be fooled by its duplex appearance, it will open the circuit in case of a fault.

Also available is a plug-in, portable kind of GFI. It gives an extra measure of safety with appliances. For example, if you are using a power drill, you can place the GFI into the wall socket first. If a fault develops, you won't get nailed. This GFI is an indoor type, but, since

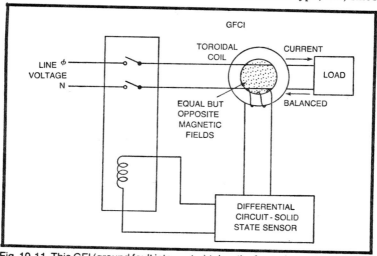

Fig. 10-11. This GFI (ground fault interrupter) takes the form of a receptacle. The appliance or lamp gets plugged into the GFI. If the balance between the current flows is disturbed, the sensing coil opens the GFI.

it is portable, you can use it in outdoor locations and bring it back into the house when you're finished.

GFI's also are available which look like ordinary circuit breakers. A GFI *is* a circuit breaker, but is also comes equipped with a built-in sensing coil. It is a super circuit breaker.

While it looks like a CB, the GFI won't plug into any ordinary panel. It will only attach to a panel that is especially equipped for it.

ATTACHING THE GROUNDING CONDUCTOR

The grounding conductor is attached to the neutral bus bar. On this bus both the system and equipment grounds are combined. The grounding conductor is common to both kinds of ground and runs from the bus bar to the electrode in the earth.

In a cable type of installation, the equipment ground is run throughout with green or bare wire. The equipment grounding wire is attached directly to the bus bar. A bonding jumper is also connected from the bus bar to the metal panel—to insure that the installation is grounded in more than one way.

In a conduit or armor type of installation, the entire metal raceway system is attached together to make all of the equipment safe and grounded.

The grounding conductor runs from the service entrance to the grounding electrode. In the city the ground wire is clamped onto the vast underground cold water pipe system. It isn't joined to the hot water pipe, since the hot water pipe does not run directly to the earth.

The grounding conductor is attached to the electrode with a pressure connector. Soldered connections are not permitted. Locknuts should be used on the connector so that all of the mating metal parts bite solidly into each other.

The parts that clamp into one another must be of the same kind of metal. Select steel clamps for steel pipes, copper clamps for copper pipes, and so on. If two dissimilar metals are clamped together, electrolytic action takes place and causes high-resistance corrosion. This could result in the pipe developing such a high resistance that the connection would become ineffective. Plastic pipe can never be used as a ground electrode.

SIZING THE GROUNDING WIRE

When you employ the cold water pipe as a ground, you should use #8 or thicker wire. If your electrode is a rod in the ground (called a "made electrode"), then #6 wire is the thinnest permitted.

If you use #8 wire as the ground wire, then it must be enclosed in conduit or armor. If you use #6 wire, the Code feels it is strong enough to be run without metal tubing to protect it.

When the location raises the possibility of the ground wire being damaged, then metal tubing should be installed. In harmless situations, though, it is cheaper to dispense with the pipe, but you should increase the size of the wire to #6 even though you are permitted #8.

Ground wire is usually bare, but there is no objection to using insulated wires.

Table 250-94 of the Code book specifies the required sizes of grounding electrode conductors in both copper and aluminum. Simply find the size of the largest service entrance conductor in copper or aluminum on the left-hand side of the chart and read off the size of the grounding conductor in one of the right-hand columns.

WATER METERS

When there is a water meter between the place you clamp onto the water pipe and the street main, you must extend your ground wire farther. The water meter is a resistance between your connection and the vast underground network that ties your installation into the earth.

You must install a jumper around the water meter. If an excessive flow of ground current should occur, the current will safely travel around the meter and not through it.

The Code doesn't directly discuss the bypassing of water meters, but it does say that "effective bonding shall be provided around insulated joints and sections and *around any equipment.*" To effect this bonding, a specially made jumper with two ground clamps is used. The jumper has a lower resistance than the meter and any current will take the jumper path rather than travel through the meter.

The jumper is only an electrical path. The water still flows through the meter.

MAKING AN ELECTRODE

An electrician thinks of an electrode as a pipe or rod driven into the ground. Commonly used as an electrode is ½-inch solid copper rod, with a length of at least 8 feet.

This kind of rod is chosen because it has a resistance to ground of less than 25 ohms. The Code book specifies that an electrode is to have a resistance of 25 ohms or less.

The Underwriters have approved a substitute rod called *copperweld*. It is made of steel with a sheathing of copper welded to the exterior.

Conduit can be used as an electrode if it is galvanized and has a trade size of at least ¾". Solid galvanized iron rods can also be used but must be at least ⅝" in diameter. The Code also lists nonferrous rods with a minimum diameter of ½" among the materials permissible for use as electrodes.

AVOIDING ELECTRODE COMPLICATIONS

In discussing electrodes the Code makes the following points. First, where practicable set the electrode into the permanent moisture zone of the soil. This could be a foot or more below the surface.

Should you encounter bedrock as you drive the electrode vertically into the earth, change your plan. If the rod is at least 8 feet long, you can remove it and bury it in a shallow trench—with the Code's approval. Be sure that your electrode is free of paint, enamel or any other nonconductive coating.

When in doubt about the resistance of an electrode, use two or more rods and bond them together so that they constitute a parallel circuit, not a series circuit.

Even though you won't be using the electrode you fabricated as a lightning rod, it is a good idea to bond all the rods together in parallel. It will limit the amount of voltage that might build up between electrodes due to capacitance effects.

When a water pipe is less than 10 feet long, make your own electrode and bond it with the water pipe.

When a building has a metal frame and is already grounded effectively, you can use the building as your grounding electrode. Any underground metal system encased in concrete (such as the reinforcing rod network in walks and driveways) can be utilized as an electrode if it answers the Code's specifications.

Underground gas pipes and storage tanks might also be satisfactory as grounds. Check with the Code and the local authority. Chances are, though, that your request will be rejected.

The ground is important. You must be serious about it. When the ground is correct, your electrical installation will be much safer than without one or with an ineffective one.

GENERAL CIRCUITS

This chapter will discuss the specifics of circuitry. Series and parallel circuits will be examined in depth. Formulas for calculating circuit resistances will be shown in practical application. Alternating current and two- and three-wire systems will be explained. The final section of the chapter will address itself to the description and comparison of single-phase and three-phase circuits.

SERIES AND PARALLEL CIRCUITS

The overhead light circuit discussed in Chapter 4 (Fig. 4-1) is a basic circuit. It has a supply voltage, switch, conductors, and a load. Is it a series or parallel circuit?

The four parts of the circuit are all in series with each other. Does that make it a series circuit? No, it does not.

The supply voltage and load are in parallel with each other. Does that make it a parallel circuit? Again the answer is "No."

It is neither a series nor a parallel circuit. That is because the terms refer to the load. It is just a basic circuit. To be in series or to be in parallel, a circuit must have more than a single load.

When a circuit has two or more loads connected in tandem, then it is a series circuit (Fig. 11-1). When a circuit has two or more loads connected across each other, then it is a parallel circuit (Fig. 11-2).

Often a single circuit has many loads, some connected in tandem and some across each other. This kind of circuit is called a series-parallel circuit (Fig. 11-3).

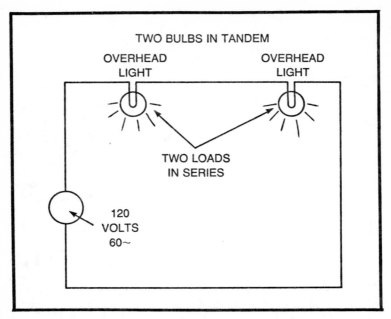

TWO BULBS IN TANDEM

OVERHEAD
LIGHT

OVERHEAD
LIGHT

TWO LOADS
IN SERIES

120
VOLTS
60~

Fig. 11-1. When a circuit has more than one load connected to it in tandem, it is called a series circuit.

Most household branch circuits, if they are not individual load circuits, are of the parallel type. They are also called "multiple circuits" and "shunt circuits."

Series circuits are not common in the electrical power schemes of North American utility companies. You'll encounter series circuits in street lighting but seldom elsewhere. You'll use series circuits for the Christmas tree lights maybe and for some TV's and radios, but not for much else.

A CLOSER LOOK AT SERIES CIRCUITS

In a lighting series circuit, all of the bulbs must be of the same size. If one is smaller than the rest, it will burn brighter than it would normally. If one is larger than the rest, it will burn dimmer than it would normally. Why?

In a typical lighting series circuit, the load (a string of Christmas tree lights, for example) is plugged into the wall receptacle where it contacts 120 volts.

The current flows into a hot line, passes through the bulbs, then returns to the wall through the neutral wire. The current remains

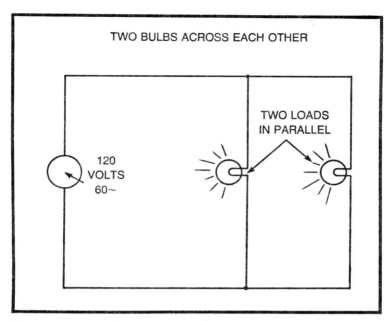

Fig. 11-2. A circuit with two or more loads connected across one another is called a parallel circuit.

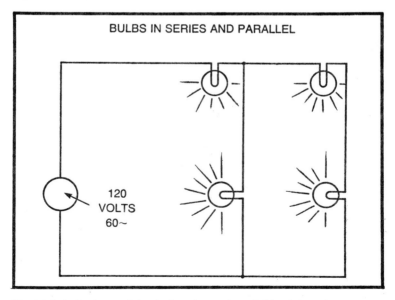

Fig. 11-3. A circuit containing both series and parallel loads is called a series-parallel circuit.

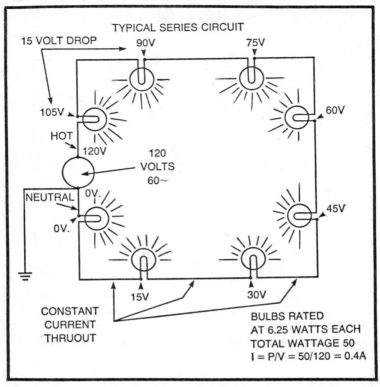

Fig. 11-4. In a typical series circuit the voltage drops from load to load while the current remains the same throughout the circuit.

the same throughout the circuit. Just as much current flows in the neutral as in the hot line.

The voltage, though, is 120 volts in the hot line and zero volts in the grounded neutral wire. The voltage drops from 120 volts to zero volts across the total load. In this instance the total load consists of eight Christmas tree bulbs (Fig. 11-4).

You can consider all eight bulbs as a single load in this series circuit. The string of lights should be regarded as one large bulb rated at 50 watts and 120 volts.

That is exactly what the total rating of all the bulbs is. Each bulb is rated at 15 volts. The total (15 × 8) voltage of the bulbs is 120. The total (6.25 × 8) wattage is 50. (See Fig. 11-5.)

As the voltage drops from 120 volts at bulb number 1 to zero volts at bulb number 8, each bulb individually drops the voltage by 15 volts.

If one of the bulbs had been rated at 10 volts, it would have been forced to drop about 15 volts, and if would have burned brighter than the 15-volt bulbs. It would behave in that manner because it was receiving more current than it should have.

Even if one of the bulbs in the string had been rated at 20 volts, it would have dropped only 15 volts. The bulb would not have been able to draw enough current to behave normally. It would have shone dimly.

PROBLEMS WITH SERIES CIRCUITS

Series circuits are not used in lighting jobs very often because each bulb is able to act as a switch.

It is possible to wire a large room with two or more overhead lights in series. When you enter the room and turn on the switch, all of the lights will go on. As you leave the room and turn off the switch, all of the lights will go off. As long as everything is intact, the circuit is satisfactory.

However, when one of the bulbs burns out, all of the bulbs will go off. If there are fifty lights in the ceiling, such as in a large commercial dining room, all fifty lights will go dark. Which one is bad?

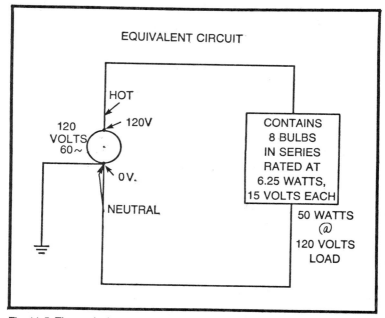

Fig. 11-5. The equivalent circuit shows the total load to be 50 watts / 120 volts.

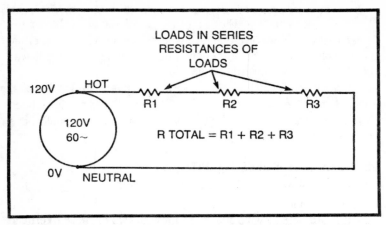

Fig. 11-6. When loads are tied together in series, the current must travel through each load in turn. The total resistance is derived by adding together the resistances of the individual loads.

You must go from bulb to bulb, testing each one in the dark, until you find the bad one. Of course it is ridiculous.

The bulbs must be prorated so that the total of the individual voltages is 120. If you divide 120 volts by 50 (bulbs), you will find that each bulb operates at 2.4 volts. This makes it impossible for the room to obtain sufficient illumination.

The appropriate way to wire most jobs is to place the bulbs in parallel. Then, instead of all the bulbs being treated by the supply voltage as a single load (as in series wiring), each bulb will be treated as an individual load.

Each bulb as an individual load can burn out and not affect any other individual load. Moreover, the rest of the bulbs will continue to shine with full 120-volt brightness, and happily the delinquent bulb can be identified immediately.

THE PARALLEL CIRCUIT

You can see that when you hook each bulb in tandem in a series circuit that the current must start a number one bulb and pass through the filament of each bulb in sequence. All of the filamentary resistances must be added together. R TOTAL = $R_1 + R_2 + R_3$....(See Fig. 11-6.)

As you add together the resistances of individual loads in a parallel circuit, the resistance decreases (Fig. 11-7). Why? As each bulb is attached across the previous one, the situation is the same as if you were to bind together the ends of a number of strands of wire. The bundle of strands would act as though they were a single piece of wire.

The cross-sectional area of each wire would add together. As the sum of the areas increased, the resistance would decrease. In effect this would produce a conductor of large gage.

As each bulb is added to the circuit, the resistance of the total load of the circuit decreases. As the load resistance gets smaller, more and more current is drawn from the supply voltage.

Each bulb (or appliance) draws a certain amount of current according to its particular resistance.

As to the voltage, each load is treated as a separate load and responds to the same amount of voltage pressure. In a typical branch circuit, the 120 volts of the hot line drops to zero volts in the neutral across each load. Each load is connected across the hot line and neutral in the 120-volt branch circuits, a practical arrangement in most instances.

CURRENT IN THE SHUNT CIRCUIT

The typical shunt circuit is a multiple 20-amp outlet circuit. A 20-amp circuit originates at a 20-amp fuse or circuit breaker. The black hot line attaches to the pole of the overcurrent device and connects to the house outlets (Fig. 11-8).

The white neutral wire returns the circuit to the panel's bus bar, not to the fuse or circuit breaker.

The outlets are attached across the two lines. The hot line is attached to the brass colored screw and the neutral is attached to the white colored screw.

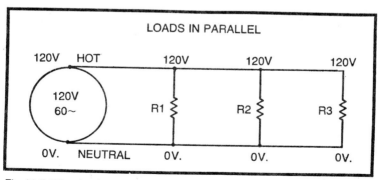

Fig. 11-7. When loads are connected in parallel, the current is divided among the loads.

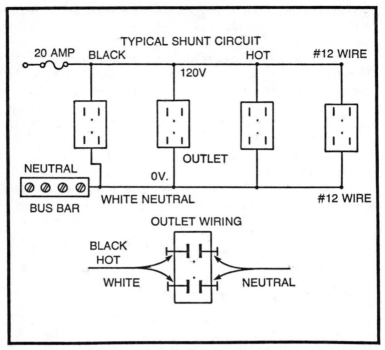

Fig. 11-8. The typical shunt or parallel circuit can be a 20-amp, 120-volt line with four duplex outlets.

The 20-amp fuse allows 20 amps to flow uninterruptedly. If more than 20 amps flow, the fuse opens.

Normally, #12 wire is used in 20-amp circuits. This size of wire carries 20 amps easily. Outlets of 20-amp rating are used in 20-amp circuits. These outlets can carry 20 amps easily.

Twenty amps of current can be drawn by a single outlet or by four outlets (as in Fig. 11-8). No more than 20 amps are permitted to flow in a 20-amp circuit. The 20 amps of current will flow in the hot line and the neutral at the same time. This is an important fact to remember. The current has the same strength everywhere in the circuit—in the neutral and in the hot line, and at every switch, outlet and load. (It's the voltage that changes: the 120 volts in the hot line drops to zero volts in the load and remains at this potential as the current passes along the neutral.)

Each outlet can be considered a separate circuit. Each outlet gets its own current. The current behaves as if each outlet were the only outlet in the circuit, even though the outlet is attached to a common hot line and neutral. Each outlet is completely independent of the others in the branch circuit.

All of the outlets share the same 20-amp fuse and black and white wires. All of the current that passes through the outlets also passes through the fuse and wiring.

Suppose there are four outlets in the 20-amp circuit. You can determine how much current will pass through the fuse by totaling the number of amps that each outlet is using. Perhaps there is a 300-watt TV on one outlet, a lamp with two 100-watt bulbs on another outlet, a 1000-watt food warmer on the third outlet and a 50-watt electric clock on the fourth outlet (Fig. 11-9). If all of the loads were operating at the same time, the total wattage would be:

(P_1)	TV set	300 W	-	2.5 A	(I_1)
(P_2)	Lamp 2-100	200 W	-	1.6 A	(I_2)
(P_3)	Warmer	1000 W	-	8.3 A	(I_3)
(P_4)	Clock	50 W	-	0.4 A	(I_4)
(P_T)	Total	1550W			

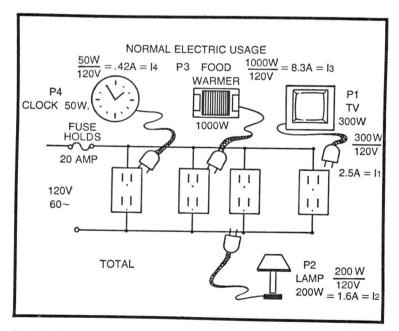

Fig. 11-9. The amount of current passing through the fuse can be calculated by totaling the number of amps that each outlet is using.

The current is $I = P/E = 1550W/120V = 12.9A$. Even if they were all on at once, the 20-amp fuse wouldn't blow.

UNLIKELY CURRENT IN THE CIRCUIT

If you should decide one day to plug in your 1800-watt barbecue, the 1000-watt food warmer, the 1200-watt electric iron, and the 1500-watt room heater all at the same time, what would happen?

$(I = PE)$					
(P_1)	Barbecue	1800 W	-	15.0 A	(I_1)
(P_2)	Warmer	1000 W	-	8.3 A	(I_2)
(P_3)	Iron	1200 W	-	10.0 A	(I_3)
(P_4)	Heater	1500 W	-	12.5 A	(I_4)
(P_T)	Totals	5500 W	-	45.8 A	(I_T)

The fuse would blow! It is only a 20-amp circuit. Why, then, are you allowed to have receptacles for four 20-amp circuits?

First of all, there is no danger because the fuse will interrupt the circuit the instant too much current is drawn. Secondly, it is highly unlikely that anyone would overload a circuit in this way. This kind of overloading is done in slap stick comedy routines, but not by real people very often.

Therefore, you are allowed to install a number of convenience outlets under the presumption that you won't be foolish enough to create an excessive load by operating heavy-drawing appliances in all of the receptacles at the same time. The receptacles are located for spatial convenience but you are expected to be prudent in their use.

Just to be safe you could wire the four outlets with an 80-amp fuse and #4 wire. That could handle almost any conceivable load, but it would be like building your driveway to the specifications of an interstate highway.

The point is that it is highly unlikely you'll ever need more than a capacity of 20 amps in the ordinary convenience outlet. You'll hardly ever operate all of the outlets at one time, much less with 20-amp loads (2400 watts).

Later in the book you'll learn about "demanding" circuits. The rationale for these circuits is that only a small portion of an ordinary wiring installation is utilized at one time. Therefore, you can apparently skimp on wiring. More about that later.

RESISTANCE IN THE MULTIPLE CIRCUIT

You should be able to understand the formula I_T (Total Circuit Current)$= I_1 + I_2 + I_3 + I_4$. I_1 through I_4 is the current drawn by each outlet (Fig. 11-10).

The formula means that the more loads you plug into the outlets, the more current you draw.

To find the resistance of a load, the Power Formula is used.

The wattage is usually imprinted on the bulb or nameplate of an appliance. The voltage of the circuit is known.

For example, what is the resistance of a 75-watt light bulb? Using the Power Formula of Chapter 4:

$$R = \frac{E^2}{P} = \frac{(120)^2}{75} = 192 \text{ ohms}$$

That is the resistance of a 75-watt bulb.

Take another example, what is the resistance of an 1175-watt waffle iron?

$$R = \frac{E^2}{P} = \frac{(120)^2}{1175} = 12 \text{ ohms}$$

That's all there is to calculating the resistance of a load. That was not too complicated, was it. Refer to Ohm's Law-Power Formula Wheel.

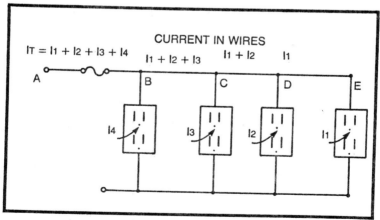

Fig. 11-10. The current in wire length AB consists of all four loads. In BC the current is all the loads except the clock. In CD only the lamp and TV currents travel. In wire length DC only the TV load current is present.

JOINT RESISTANCE

When there is a single load in a circuit, you can calculate the current consumption in amps and its resistance in ohms quickly. As additional loads are plugged in, you can calculate each in turn. That is easy, but what about their total resistance? The total resistance of the loads in a circuit is the sum of the individual resistances. It is called "joint resistance."

While the currents and wattages simply add together and cause an increase in amps and watts in the circuit, the resistance decreases.

As each additional load is plugged in, the resistance takes a drop each time. It is like twisting a number of wires together in parallel. You are increasing the total cross section of the copper and giving the current a broader avenue for flow. The joint resistance (expressed in ohms) keeps decreasing.

Memorize the rule, the total joint resistance in multiple circuits is always LESS than the value of the smallest individual resistance.

The formula is:

$$R_1 \text{ (Total Joint Resistance)} = \frac{1}{1/R_1 + 1/R_2 + 1/R_3 + 1/R_4}$$

Let's calculate the joint resistance of the 75-watt light bulb and the 1175-watt waffle iron of the previous section. You'll remember that their resistances were 12 ohms and 192 ohms, respectively.

$$R_T = \frac{1}{1/R_1 + 1/R_2}$$

$$= \frac{1}{1/12 + 1/192}$$

$$= 11.3 \text{ ohms}$$

Notice that the answer is 11.3 ohms, a quantity which is smaller than the lesser (12 ohms) of the two individual resistances. (The calculations are easily done with a pocket calculator. Every electrician should own a pocket calculator. It'll be one of his important tools.)

ALTERNATING CURRENT

It was mentioned earlier that an electric generator is nothing more than a magnet and conductor is close proximity, with one (either) of the components in rotational motion. A 60-cycle generator consists of a wire coil which rotates inside a magnetic field at 60 rotations per second.

As the coil rotates, the electrons in the magnetic field are attracted and repelled. The electrons move first in one direction, then slow down until they gradually halt completely. Then they reverse direction, attain their original speed, and gradually slow to a full stop. This two-step process represents one cycle (Fig. 11-11). In recent years the hertz has been adopted as the unit for expressing cycles per second.

The movement of the electrons constitutes electrical energy. It doesn't matter much whether the electrons move in a single direction or alternate back and forth. One-direction movement produces

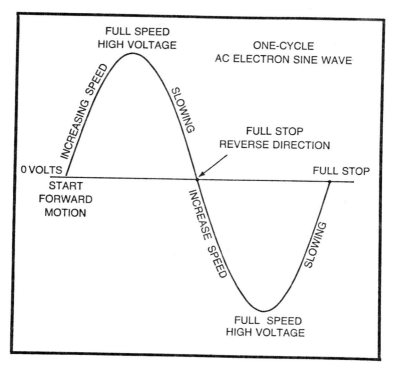

Fig. 11-11. The movement of electrons in a wire can be graphed. One cycle of AC looks like this.

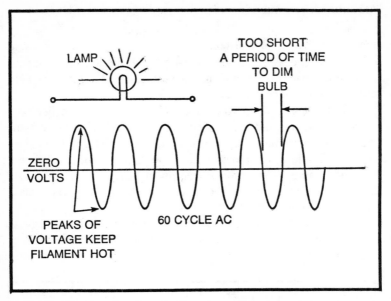

Fig. 11-12. The 60-cycle voltage is able to keep the filament of a lamp hot, even between voltage peaks.

direct current (DC) and two-direction movement produces alternating current (AC). The important thing is that the electrons flow and as they flow carry electrical impulses through the conductors to the loads. The impulse travels at the speed of light, 186,000 miles per second.

The impulse causes the electrons in the lamps or appliances to move. As the electrons move in a heating element, the element gets hot. As the electrons move in the filament of a bulb, the filament glows white hot and throws off light.

As the electrons accomplish the back and forth cycle, they reach maximum acceleration twice. They also stop dead in their tracks (as they reverse direction) twice each cycle.

While the electrons are at maximum speed, they create maximum heat and magnetic fields. While they are at rest (midway through each cycle), no energy is expended. In between maximum speed and rest, more or less energy is created, according to the rate of movement of the electrons.

The white filament of the bulb does not have time to cool off as the electrons slow down and stop. The next pulse arrives soon enough to prevent the filament from cooling (Fig. 11-12).

A motor connected to the circuit does not have time to stop rotating between the magnetic pushes it receives while the electrons are at maximum speed (Fig. 11-13).

Thus the 60-cycle AC is satisfactory to do our electrical chores for us. Alternating current has a number of advantages over direct current and is the preferred system. Most of our homes use single-phase, three-wire electric systems. In rare cases single-phase, two-wire systems are used. Let's examine each of these systems.

THE THREE-WIRE SYSTEM

The commonest type of electric service in the U.S. is 120/240 volt, 60 cycle. This has been the standard service for a long time and most American lights and appliances are designed to be energized with this kind of electric power.

The three-wire system receives its name from the fact that two hot lines and one neutral wire are used to bring the current to the service entrance.

Actually what is being brought into the building is 240 volts, 60 cycle. That big "pot" on the electric pole outside is a transformer. One of the transformer's windings issues 240-volt current. A current of 240 volts and 60 cycles courses back and forth in the winding (Fig. 11-14).

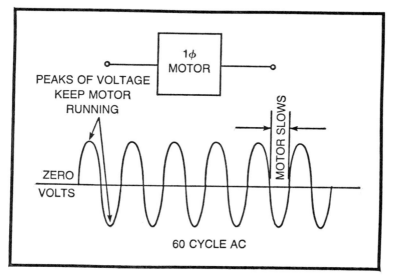

Fig. 11-13. At every peak in either direction a motor receives a magnetic shove which keeps it rotating.

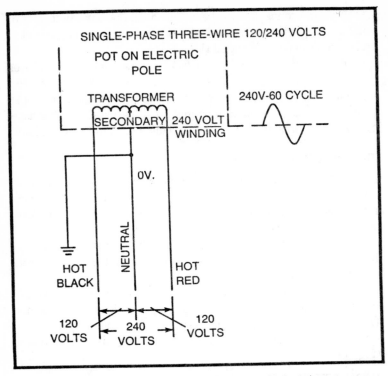

Fig. 11-14. The single-phase, 3-wire distribution system is one of the most used electrical home supplies.

A wire is connected to each end of the winding. These wires are the hot lines. Typically one wire is black and the other is red. Across the two wires are 240 volts potential, alternating current. When the black wire is at 240 volts, the red wire is at zero volts.

As a half of a cycle passes, the black wire goes to zero volts and the red wire to 240 volts. The system is like a see-saw. When one side is up, the other side is down. It is the same current, just different ends of the current.

Where does the 120 volts come from? The 120 volts comes from the midpoint between the two ends of the winding, the dead center of the winding. When the black wire is at 240 volts, the center of the winding is at 120 volts, and the red wire is at zero volts.

Black and red are both 120 volts away from the center of the winding. When the current reverses and red becomes 240 volts, the center is at 120 volts and black is at zero volts.

A third wire is attached to the center of the winding. It is always 120 volts away from the black and red windings.

The next step is to ground the center tap and call it neutral. The ground makes the center tap lock in at zero volts. As the AC peaks and nulls, the red and black wires remain 120 volts away from the center.

As a result, we have a three-wire system with 240 volts AC potential between the black and red wires, 120 volts AC potential between black and neutral, and 120 volts AC potential between the red and neutral.

THE TWO-WIRE SYSTEM

A typical two-wire system can only provide 120 volts. The same type of winding is in the transformer of the pot, except that across the winding is 120 volts (Fig. 11-15).

A black wire and a white wire are used to carry the voltage. The white wire is grounded. This locks that end of the winding to zero volts. The black wire is attached to the other end of the transformer's winding and carries 120 volts.

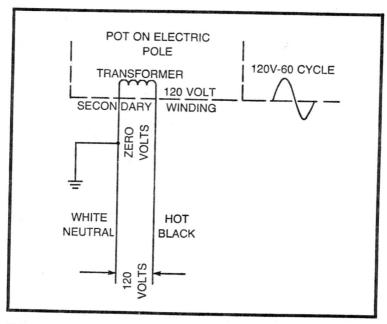

Fig. 11-15. The single-phase, 2-wire distribution system is rarely used today. It does not provide 240 volts.

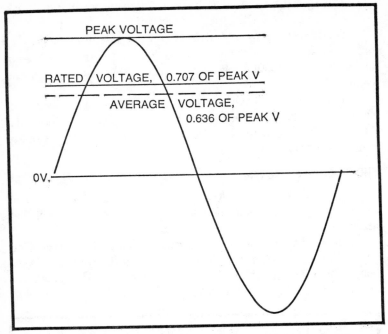

PEAK VOLTAGE

RATED VOLTAGE, 0.707 OF PEAK V

AVERAGE VOLTAGE,
0.636 OF PEAK V

0V.

Fig. 11-16. The rated voltage is the one that is measured with a meter. The peak voltage is only present for an instant each cycle.

You can only use 120-volt (also 115-, 117-, or 110-volt) lamps and appliances on the 120-volt system. You cannot use a 240-volt lamp or appliance.

The 120 volts is called the *rated* voltage. This is because there is also a *peak* voltage. The peak voltage is about 170 volts.

If you graph the motion of an alternating current during one cycle of 120 volts, you'll observe the following. The voltage begins at zero volts and proceeds to increase. It rises to a peak and then decreases to zero. The load utilizes the rated (or effective) voltage, not just the peak voltage. The peak is only present for an instant. The effective voltage approximates the average 120 volts and represents the amount of voltage actually being used (Fig. 11-16).

Mathematically, the useful voltage is considered to be 0.707 of the peak voltage.

THREE-WIRE ALLOWS BALANCING

At first glance, a three-wire system appears to be nothing more than two two-wire systems with their neutrals hooked together

(Fig. 11-17). While you could produce 240 volts from this circuit arrangement, it is a pseudo three-wire system. Besides other complications, there would be two neutrals (making it some sort of four-wire system). Each neutral would be independent of the other. Current would flow in both neutrals at full strength and the neutral wire would require a diameter equal to that of the hot lines.

Each side of a three-wire system is able to energize 120-volt loads. However, the two sides share the neutral (Fig. 11-18). When current is flowing in one direction on one side, it is flowing in the opposite direction on the other side.

This means that two currents (one from each side) are traveling in the neutral, but in opposite directions. The two currents cancel each other. This eliminates heat and magnetic energy (which are a waste, anyway) in the neutral. Also, since there is less current in the neutral, a wire of smaller size can be used.

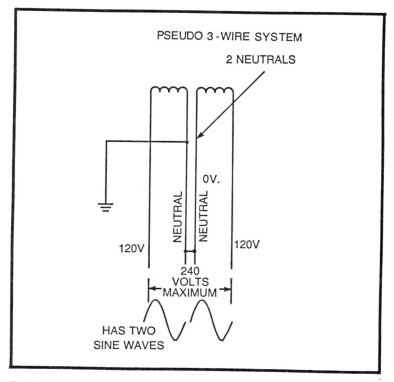

Fig. 11-17. Two 2-wire distribution systems can produce 120/240 volts but not with efficiency.

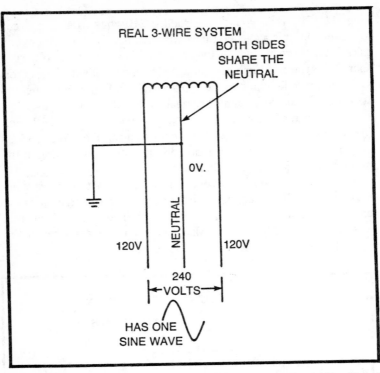

Fig. 11-18. A real 3-wire system has 2-wire distribution systems at 120 volts and one 2-wire system at 240 volts.

How does this work? Take the example of a three-wire entrance in normal operation. If the 120-volt branch circuits are equally distributed between the two 120-volt sides (Fig. 11-19), the following currents could be drawn during operation.

Side A	12 amps
Side B	15 amps

If the two currents are opposite in direction and are returned to ground in the neutral, then the neutral ends up with the amount of current that is not canceled: 3 amps, the difference between side B and side A.

The uncanceled flow is called the unbalanced current. In a two-wire system you can't balance currents. The neutral carries just as much current as the hot line at all times.

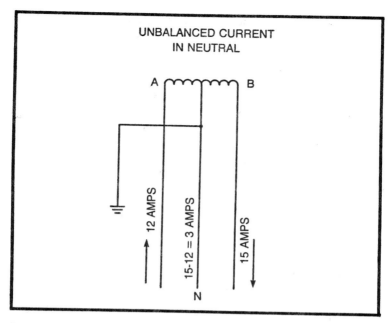

Fig. 11-19. A real 3-wire system can be connected so that the currents in the two sides can be balanced as closely as possible.

THREE-WIRE PRECAUTIONS

When you wire up the loads of a three-wire circuit, great care must be taken to put just as much load on one of the 120-volt sections as on the other. For example, if you have two 20-amp small appliance

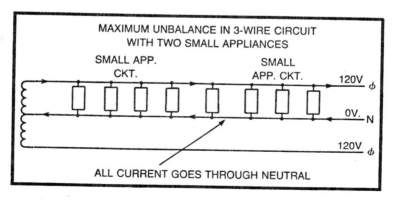

Fig. 11-20. Should a 3-wire system be connected without balancing, the neutral wire would carry too much current.

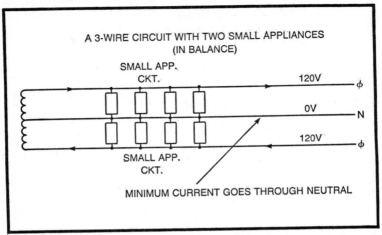

Fig. 11-21. When a 3-wire system has its loads balanced correctly, minimum current passes through neutral.

circuits in a home, one of the circuits must go on one 120-volt line and the other circuit on the other side. What would happen if you put both circuits on the same side? (See Figs. 11-20 and 11-21.)

If both circuits were connected to one side, then the current from each would travel in the same direction in the neutral. The two small appliance currents would add together. There would be more current in the neutral than in either appliance line. This would be dangerous.

The neutral wire would get hot. Since there is never an over-current device in a neutral line, the heating would continue until a fire started.

A second worry in three-wire circuits is the continuity of the neutral. The neutral must never be open when the circuit is active. It must present a continuous electrical path for the current from each side of the 240-volt lines. If a neutral should become disconnected, a lot of loads will stop being in a 120-volt parallel circuit and become involved in a 240-volt series circuit (Fig. 11-22).

Bulbs will burn out, appliances overheat and other damages occur.

SINGLE-PHASE CURRENT

The ordinary two-wire and three-wire systems are single phase. As mentioned, the current is produced by rotating a coil of wire in a magnetic field (Fig. 11-23) or by rotating a magnetic field

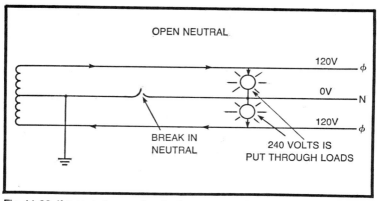

Fig. 11-22. If a neutral opens in a 3-wire system, a dangerous situation occurs. The system could become ungrounded and 240 volts could pass through 120-volt loads.

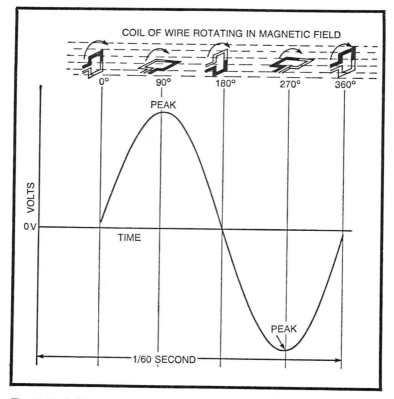

Fig. 11-23. A 60 cycle per second AC voltage is induced in the coil of wires as it rotates in the magnetic field.

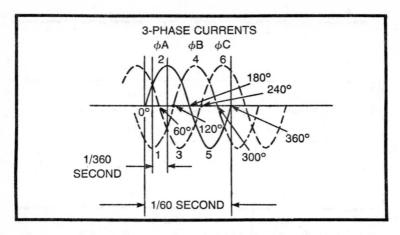

Fig. 11-24. Each separate winding on a 3-phase rotor produces an individual current.

near a coil or wire. Either way, if you graph the voltage through one cycle, you get the shape of a sine wave.

Every cycle is identical to every other cycle in this electric generator. So if you understand this one, you understand them all.

The horizontal axis of the graph is time. In this case it is 1/60 second from beginning to end. During this interval the generator makes one rotation, from 0° to 360°.

The vertical axis of the graph is voltage. Peak voltage is 170 and rated voltage is 120. At the beginning of one rotation the voltage is zero. Zero is the start of this phase.

In fact, there is only one phase in this case, since the rotor consists of a single coil. If there were three coils (spaced out) on the rotor, then three separate voltages would be produced and the current would be three-phase.

We'll examine three-phase current next. Meanwhile, it can be seen that as the single coil rotates, its electrons are set into motion by the effect of the magnetic field. As you plot the voltage on a graph, you trace a line from zero to peak, back to zero, to peak in the opposite direction, and back to zero.

THREE-PHASE CURRENT

Whereas one coil produces single-phase current, three coils on the rotor of a generator produce three-phase current. The three coils are equally spaced on the rotor at 120-degree intervals.

As the rotor revolves at 60 hertz, three phases of current are produced, each at a slightly different time. Let's refer to them as phases A, B and C.

Phase A peaks first, then B, then C. The peaks are staggered neatly. A graph of the three sine waves shows that a sine wave peak is occurring every 1/360 second (Fig. 11-24).

In single-phase current, a sine wave peak, top and bottom, occurs every 1/120 second.

Why do we even need three-phase current? Isn't single-phase good enough. So what if there are six peaks per cycle instead of two?

Three-phase current is very important to the operation of electric motors. Single-phase current gives the motor two magnetic pushes per second, one push at each peak. Between pushes the motor tends to slow down. A heavy duty motor does in fact slow appreciably with single-phase power.

If three-phase current is used, the motor gets six pushes every second. This keeps it rotating smoothly and easily. It needs a more consistent impetus than single-phase current can provide.

Another example of a load requiring three-phase power is a heavy duty commercial oven. The oven has trouble getting hot enough with single-phase power. The elements tend to cool between the peaks. Six peaks is almost continuous power, instead of the voltage fluctuating widely between zero and 170 volts. With six peaks the temperature drops, but never more than a bit.

KINDS OF THREE-PHASE CURRENT

There are two common types of three-phase current. One is called "delta" and the other is called "wye." The delta is named after

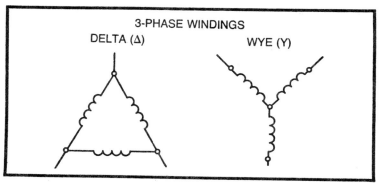

Fig. 11-25. The two common types of 3-phase windings are called delta and wye.

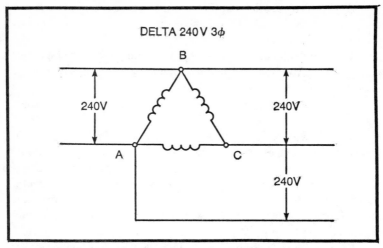

Fig. 11-26. The delta can deliver three separate 240-volt currents to power 3 loads.

the Greek letter △ because a schematic drawing of the windings resembles a delta. Wye is the word form of the letter Y and is descriptive of the winding's configuration (Fig. 11-25).

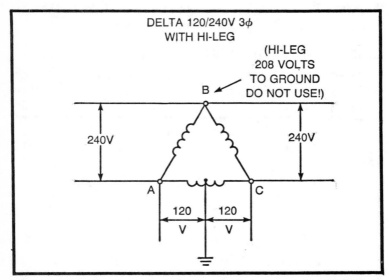

Fig. 11-27. The delta when one winding is center tapped can deliver 120/240 volt 3 φ. Opposite to the center tap is the HI-LEG. It is 208 volts to ground. Do not use it to power 120-volt loads.

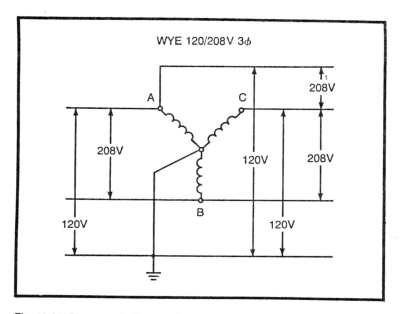

Fig. 11-28. The wye winding can deliver 120/208 volts 3 φ.

Across each delta winding ordinarily there are 240 volts. The three windings are independent of each oher. Each 240-volt sine wave can be used on its own. Each winding can power a 240-volt circuit (Fig. 11-26).

If you ground the center of a delta winding, you have a three-wire, 120/240 volt system. It can power a residence like any 120/240 volt system (Fig. 11-27).

The Y system is called 120/208-volt, three-phase. From any one of the legs to ground is 120 volts, like a two-wire, 120-volt system. From leg to leg is 208 volts (Fig. 11-28). More about this in the next chapter.

WIRING CIRCUITS

This chapter is concerned with the circuit connections. After the right size pipe, wire, fittings, and devices are installed into components, and grounded, then the circuits have to be hooked up.

CIRCUIT HOOKUP

You must make all your wire connections so that electricity will flow and energize the loads in exact accordance with the design of the system. The word exact means just that. There is no almost in circuit hookups.

Maybe when you installed the pipe, there was a place where a strap was impossible to install. The pipe was secure and eliminating the one strap didn't adversely affect the overall installation a bit.

When you connect a circuit, there is no such leeway. A connection cannot be left off or placed on the wrong terminal. It must be connected correctly. If it is not, the circuit won't work properly. A motor won't start, a bank of fluorescents won't light up, or the wiring in a wall will overheat and start a fire.

To hook up a circuit, you must understand the circuit. You can follow a schematic if it is available, but most of the time you'll only have a print with architectural symbols. It is up to you, without asking anyone, to connect the circuit correctly.

THE BASIC CIRCUIT

Previous chapters have presented considerable discussion on circuits. You've seen lots of schematic diagrams and you know that black or red wires are normally the hot lines and that white or gray wires are normally the neutral lines. You also know that bare or green wires are used to ground equipment. Fortunately the industry has standarized these colors and a lot of confusion has been avoided.

Terminal screws on outlets, devices, fixtures, and other items also have standardized color tones. There are gold and brass screws to which the black wire attaches and white or silver screws to which the white wire attaches. Equipment grounding terminal screws are often painted green. All of this color coding saves the electrician a great deal of time.

The overhead light circuit which we've discussed is a typical basic circuit. To hook up such a circuit you'd need two lengths of black wire, one length of white wire, an OFF-ON switch and a light fixture (Fig. 12-1).

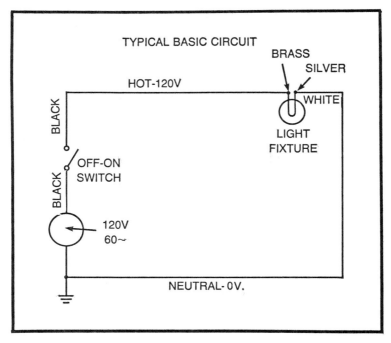

Fig. 12-1. The voltage drops from 120 to zero across the filament of the light. The current is the same in the hot line and the neutral.

One length of black wire connects the 120-volt source to one of the terminals on the OFF-ON switch. It doesn't matter which terminal is selected.

The second length of black wire connects the remaining terminal of the switch to the brass terminal of the fixture. The two black wires carry the current at 120 volts.

The fixture and the light bulb drop the 20 volts to zero volts. The white wire connects to the silver terminal of the fixture and returns the circuit, unbroken, to the neutral of the source.

This basic circuit is common. Review the way the current remains the same throughout the circuit while the voltage drops from 120 to zero.

CONNECTING THE SWITCH

In the basic circuit, the switch could have been placed anywhere along the circuit and work equally well. That is, it could have turned the circuit off and on as it opened and closed the circuit as the toggle was moved.

However, as has been mentioned, the neutral wire must be continuous and never broken. Even though the switch would operate well enough in the white wire, from a safety standpoint you should not install a switch in the white neutral. If you did, you'd produce an ungrounded neutral wire which could carry the hot line potential. Neutral is designed to remain always at zero volts, never at the hot line voltage.

The switch then always is in the hot line. It is never connected in the neutral or between the hot and neutral. It is connected in series with the hot line and should be thought of as part of the hot line.

If you have a black wire joining a piece of equipment to which you'd like to apply a switch, merely cut the black wire at any convenient place. Then trim the insulation at the tips and attach the ends to the two terminals of the switch.

Thus you'll be able to open and close the black wire at will, thereby opening and closing the circuit. The switch is always in series with the load.

The switch should not be in parallel with the load. This situation would exist if you attached a white wire to one of the terminals and a black wire to the other. Should this accidentally happen, turning on the switch would create a dead short from 120 volts to ground. The fuse would blow when the current started passing through the switch.

CONNECTING THE OUTLET

The point is being emphasized that switches go in series with the black wire, but you should realize that outlets do not. An outlet is sometimes confused with a switch during hookup. This is dangerous.

If you hook up an outlet in series with the black wire, the loads as they are plugged in will operate in series with the hot line. This could burn out the equipment.

An outlet should be installed in parallel with the supply voltage and the rest of the loads and outlets. This means that an outlet gets a black wire attached to its brass terminal and a white wire to its silver terminal. It is hooked across the 120-volt line and the zero-volt line (Fig. 12-2).

Just as the switch is considered part of the black wire, the outlet is thought of as part of the black wire and white wire. The switch and outlet connection are different and must never be confused with each other.

If an outlet has a grounding terminal, the green or bare wire should be attached to the terminal.

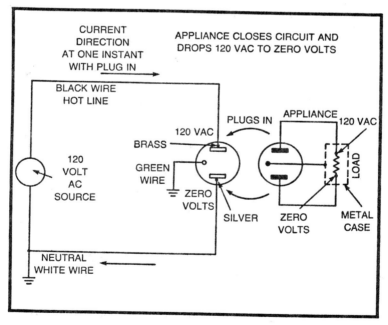

Fig. 12-2. An outlet is a passive device into which the load can conveniently be plugged.

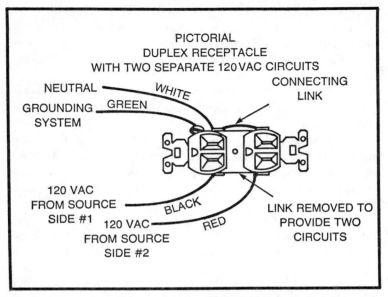

PICTORIAL
DUPLEX RECEPTACLE
WITH TWO SEPARATE 120 VAC CIRCUITS

NEUTRAL
GROUNDING
SYSTEM

WHITE
GREEN

CONNECTING
LINK

120 VAC
FROM SOURCE
SIDE #1
120 VAC
FROM SOURCE
SIDE #2

BLACK
RED

LINK REMOVED TO
PROVIDE TWO
CIRCUITS

Fig. 12-3. The typical duplex receptacle can be wired so that both outlets are in the same circuit or in separate circuits.

An outlet is a passive device. It can't contribute to any work until an appliance or equipment cord is joined to it. The outlet is simply a convenient device to plug into. It has no moving parts.

When equipment is plugged in, the current can flow from the black wire into the equipment and back out to the white wire. The voltage is 120 (or 240) when it leaves the black wire and is zero when it enters the white wire. The equipment drops the voltage through itself as it operates.

The green grounding terminal keeps the frames of the outlet and the equipment at ground zero.

DUPLEX RECEPTACLES

Most of the 15-amp and 20-amp outlets are duplex receptacles. The heavier amperage types are single outlets.

Duplex receptacles can be wired so both outlets are in the same circuit or on individual circuits. Either way there is only one neutral wire and one grounding wire needed.

The two neutral terminals are shorted together by a metal strip linking them. The white wire is attached to one of the neutral terminals. The green grounding terminal is a single terminal and is not in the circuit at all, its function being to ground the frame of the receptacle to the equipment grounding system.

The two brass terminals are the ones that determine whether the two outlets are wired together in parallel or individually. When they are wired together, a metal link like the one between the neutral terminals is left in place. Then a black wire is attached to one of the brass terminals.

If the two outlets are to be installed in separate circuits, then the connecting metal link is removed (Fig. 12-3). A black wire from either circuit is attached to one of the brass terminals. It does not matter which terminal.

Another hot line, either black or red, is then attached to the other brass terminal. This line is usually taken from the opposite side of the source so that the two circuits will balance each other (Fig. 12-4).

Both wires are 120-volt lines but as the black wire reaches its voltage peak, the red wire is at zero volts. When the red wire is at its

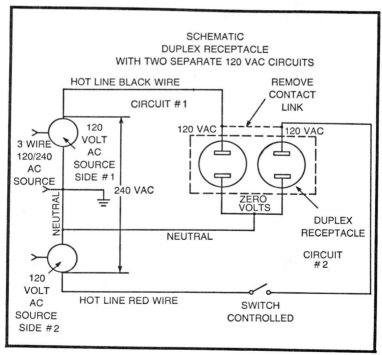

Fig. 12-4. In small appliance circuits a duplex receptacle can be useful in balancing and reducing neutral current.

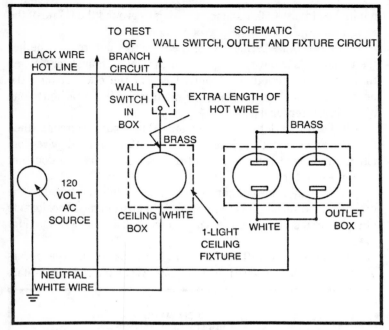

Fig. 12-5. The schematic diagram of the circuit shows the connections.

peak, the black wire is at zero volts. Thus, the neutral is common to both outlets and will have the smallest current possible.

SWITCH, OUTLET AND FIXTURE

A combination of a switch, outlet and fixture can be placed at the end of a branch circuit. The combination could consist of an overhead light controlled by a wall switch and a duplex receptacle that is always hot.

In the wiring arrangement the last thing on the line is the receptacle. Next to last is the wall switch and then the fixture.

The schematic diagram shows that the receptacle is in parallel with the supply voltage. The switch and fixture are in series with each other but in parallel with the outlet, across the hot line and neutral.

It is easily seen on the schematic that the receptacle is not switched and remains hot all the time (Fig. 12-5). The ceiling fixture is switched.

In the actual wiring (Fig. 12-6), the black and white wires enter the ceiling box from the 120-volt source. The two wires then con-

254

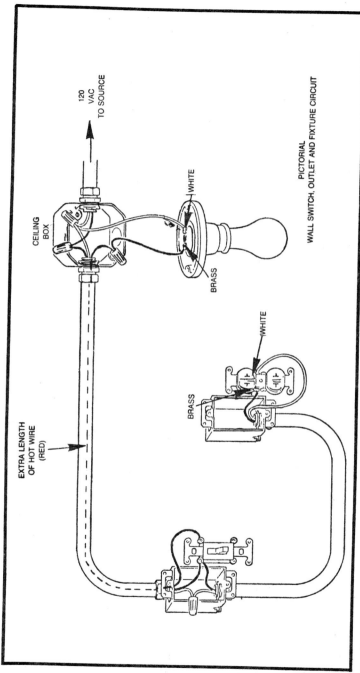

120
VAC
TO SOURCE

CEILING
BOX

WHITE

BRASS

WHITE

BRASS

EXTRA LENGTH
OF HOT WIRE
(RED)

PICTORIAL
WALL SWITCH, OUTLET AND FIXTURE CIRCUIT

Fig. 12-6. The pictorial diagram of the circuit shows the actual hardware.

tinue to the switch box and two more wires are run from the switch box to the outlet box.

A second hot line, perhaps with red insulation, is also run from the ceiling box to the switch box. You are now ready to make the hookup. The hookup has to follow the schematic. You must learn to look at the schematic and relate the schematic symbols to the actual hardware.

Connections are first made to the receptacles. Black wire goes to brass, and white wire to silver. This places 120 volts on the poles to which the black wire is attached and zero volts on the poles to which the white wire is attached. The 120-volt potential is across the open circuit at all times. As soon as a load is plugged in, the load can drop the voltage as it absorbs current.

The next connections are made to the fixture. The white wire goes to the neutral terminal and the red wire goes to the brass terminal.

The last connections are made to the switch. The red wire goes to one side of the switch. The insulation is stripped from the end of the black wire which runs past the switch. The bared end is then attached to the other end of the switch. Neutral is not, and should never be, attached to the switch. It just runs through the switch box.

THE THREE-WAY SWITCH

The symbol S_3 on the architect's drawings indicates three-way switch (Fig. 12-7). You can identify an actual three-way switch

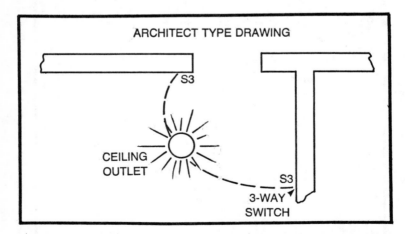

Fig. 12-7. On the architect's print the 3-way switch and circuit are shown as two S_3's and an outlet.

because it has three connections, instead of the two that are on an ordinary switch (Fig. 12-8).

Three-way switches are used in pairs. Their purpose is to offer switching control of a circuit from two locations. For example, an overhead hallway fixture could be controlled from either end of the hallway.

There are a number of possible layouts. However, if you consider the two switches as one unit, they are wired just like a single switch. They are in *series* with the load in the *hot* line (Fig. 12-9).

Inside the three-way switch is a single-pole, double-throw switch. In the circuit hookup the two ends of the hot line are attached to the single-pole connections of the two switches.

The two other terminals on the switch are then hooked together with two parallel wires. These twin wires are called "traveler wires" or "switch lines." They attach to the traveler terminals of the two three-way switches.

The switches are designed to produce the following circuit condition without fail. The toggle of each switch has to be able to turn the light on and off, no matter what the position of the toggle.

For example, when you enter the dark hallway, you should be able to flick either toggle and make the light go on. If the light is already on, movement of the toggle in either direction should turn it off.

When the circuit is closed, the light is on. When the circuit is open, the light is off. The necessary conditions are obtained by the use of the twin traveler wires. If the circuit is open, moving either switch will close the circuit. If the circuit is closed, moving either switch will open the circuit. Four possibilities are shown in Fig. 12-10.

FOUR-WAY SWITCH

The symbol S_4 on the architect's plans indicates four-way switch. You can recognize the four-way switch because it has four terminals instead of two like an ordinary switch or three like a three-way switch (Fig. 12-11).

The four-way switch is used with a pair of three-way switches for control of an additional (a third) location; for example, a patio with one overhead light and three rooms exiting onto the patio. Each room should have a switch located inside by the doorway to turn on the patio light. Two three-way switches and a four-way switch are used to accomplish this purpose.

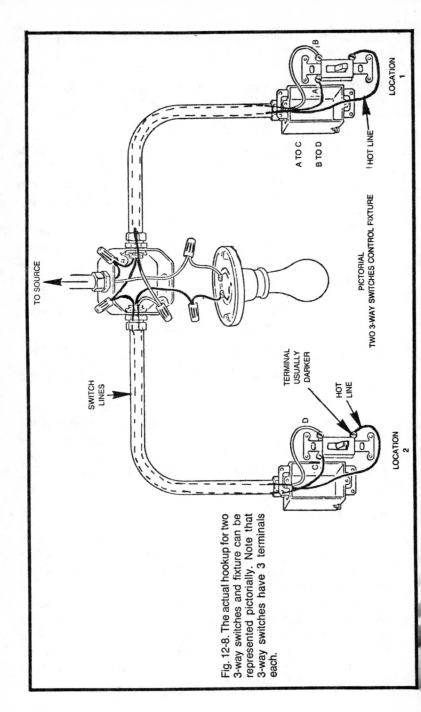

TO SOURCE

HOT LINE

LOCATION 1

A TO C

B TO D

PICTORIAL
TWO 3-WAY SWITCHES CONTROL FIXTURE

SWITCH LINES

TERMINAL USUALLY DARKER

HOT LINE

LOCATION 2

Fig. 12-8. The actual hookup for two 3-way switches and fixture can be represented pictorially. Note that 3-way switches have 3 terminals each.

Inside a four-way switch is a mechanism that alternates the two pairs of switching contacts. When the toggle is inclining in one direction, the two contacts on the two ends are connected together. When the toggle is inclining in the other direction, the two contacts connect in the other possible manner. There are no ON or OFF markings (Figs. 12-12 and 12-13).

If you want to control the light from more than three locations, more four-way switches are installed. One of the three-way switches is attached to the load end of the circuit and the other three-way switch to the supply end. In between are all of the four-way switches (Fig. 12-14).

The four-way switches when attached to the pair of three-way switches with traveler wires fulfill the requirements. Any of the three-way or four-way switches can turn the light OFF if it is ON and ON if it is OFF. Figure 12-5 shows the current flow and continuity.

When actually connecting the traveler wires, be careful that you connect the traveler wires from the three-way switch to one end of the four-way switch. The other end of the four-way switch attaches

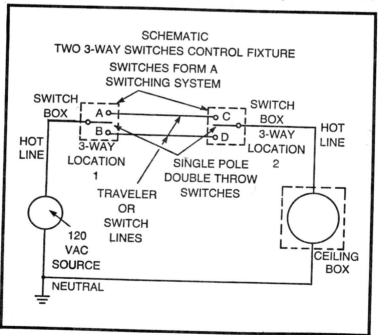

Fig. 12-9. The two 3-way switches are connected as a unit in series with the hot line.

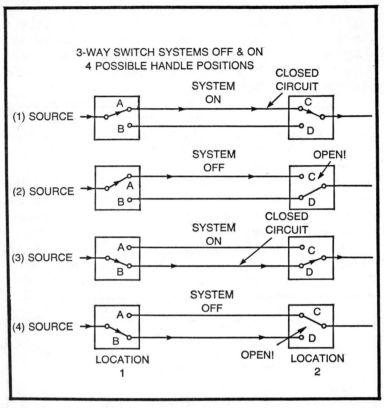

Fig. 12-10. The two 3-way switches have four possible positions, two on and two off.

to the wires of the other three-way switch. Do not connect a pair of traveler wires across the switch.

CONNECTING THE CIRCUIT BREAKER OR FUSE

The circuit breaker looks like a form of toggle switch, but it is not; it is an overcurrent device like a fuse (Fig. 8-5). It is wired into series with the hot line, the so-called ungrounded line. Every ungrounded line in a building has a circuit breaker or fuse installed in series with it. Thus, if too much current is drawn through the hot line, the device interrupts the circuit. The circuit breaker is installed in a circuit breaker panel.

In the single-phase, two-wire panel, the fuse is screwed or plugged into a socket. One end of the socket is connected to the

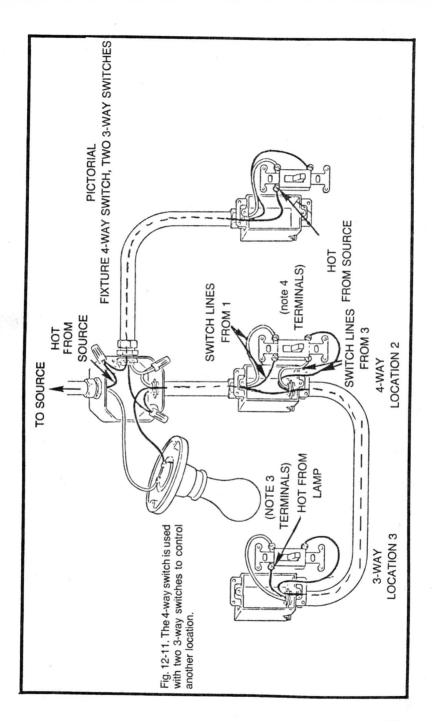

PICTORIAL
FIXTURE 4-WAY SWITCH, TWO 3-WAY SWITCHES

TO SOURCE

HOT FROM SOURCE

SWITCH LINES FROM 1

(note 4 TERMINALS)

HOT SWITCH LINES FROM SOURCE

SWITCH LINES FROM 3

4-WAY LOCATION 2

(NOTE 3 TERMINALS) HOT FROM LAMP

3-WAY LOCATION 3

Fig. 12-11. The 4-way switch is used with two 3-way switches to control another location.

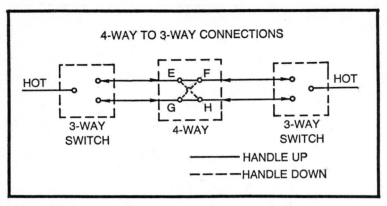

Fig. 12-12. A 4-way switch is a mechanism that alternates two pairs of switching contacts.

incoming hot line. The other end of the socket is connected to the hot line of the branch circuit.

The neutrals in and out are attached to the neutral bus bar. There is no connection between the fuse and the neutral, either in or out.

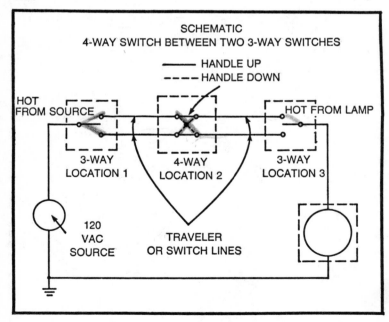

Fig. 12-13. Between two 3-way switches according to the desired number of control locations, 4-way switches are installed.

In the single-phase, three-wire panel, there are two hot bus bars and the neutral bus bar. The circuit breakers can plug into the two hot bus bars, not across them because one end of the circuit breaker attaches to one bus bar (Fig. 10-6).

All of the circuit breakers are evenly distributed between the two bus bars in a balanced installation.

The other end of the circuit breaker is attached to one of the various branch circuits. For a 120-volt branch, one circuit breaker is used. The end to the bus bar is usually plugged or bolted to the bus bar. The end to the branch circuit has a screw terminal to which the wire attaches.

If it is a 15-amp circuit breaker, a #14 wire is attached. A 20-amp circuit breaker gets a #12 wire, and so on up.

For a 240-volt branch, two poles in a dual circuit breaker arrangement are used. There is a pole for each hot line. Again let us mention that there is no connection between the neutral and the circuit breaker.

Each pole of the dual circuit breaker is connected to a different bus bar. Therefore, each circuit breaker is in series with each hot line. At no time are both poles to be connected to the same bus bar.

When the two hot poles of the 240-volt branch circuit are connected to the two bus bars, the circuit is automatically balanced. If there is a neutral return wire in the 240-volt branch circuit, there will be no current in the neutral, except for the 120-volt connections.

120/208, FOUR-WIRE, THREE-PHASE CIRCUIT BREAKERS

In this four-wire panel there are three ungrounded bus bars, each 120 volts to ground and each coursing along at a different phase.

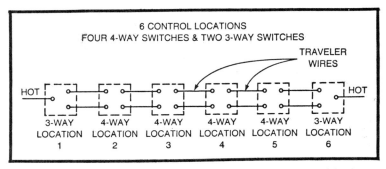

Fig. 12-14. To control a light from more than three locations, one of the 3-way switches is attached to the load end of the circuit and the other 3-way switch to the supply end. All of the 4-way switches lie between.

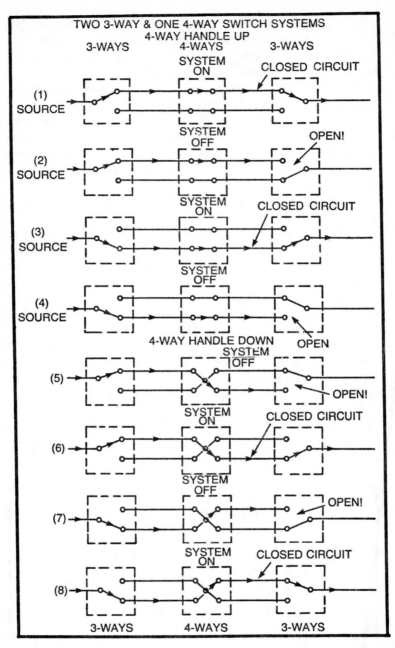

Fig. 12-15. The 4-way switches with two 3-way switches fulfill the ON-OFF requirements.

Each bus bar will provide a two-wire service of 120 volts from a hot bus bar to the neutral bus bar. Neutral is common to all three ungrounded bus bars (Fig. 12-16).

For the 120-volt branch circuits, the circuit breaker connections are the same as for the three-wire single phase. A single-pole circuit breaker is plugged into a bus bar and the branch circuit's ungrounded wire is screwed onto the other terminal of the breaker.

The 120-volt branch circuit is thus protected against overload. Should the branch circuit begin to draw too much current through the circuit breaker in series with the hot line, the circuit breaker will trip and interrupt the circuit.

The 120-volt branch circuits are equally distributed to the three hot bus bars. This balances the 120-volt branches for proper operation.

When a two-pole circuit breaker is installed, it is installed like the three-wire service. One pole each of a dual circuit breaker is

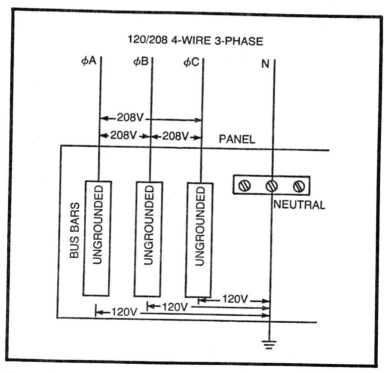

Fig. 12-16. Between any two ungrounded bus bars is 208 volts 1φ. Between any ungrounded bus bar and the neutral bus bar is 120 volts 1φ.

attached to different bus bars. There are three hot bus bars, and the dual circuit breaker is attached to any two of them.

Care is taken to balance the circuit breakers according to the loads they are to service. Instead of the circuit breaker being automatically balanced as it supplies 208 volts, as are the single-phase, 240-volt circuit breakers, the three-phase needs care in balancing.

Across any two of the three hot bus bars is 208 volts. It also acts as a single-phase 208 volts, as the 120 acts as single-phase with each individual bus bar.

There are two phases when you connect to two bus bars. The voltage of each phase is 120 volts. If you consider each phase separately, then the total of the two is 240 volts.

However, the two phases are 120 degrees out of phase with each other. Therefore the voltages add together somewhat differently than would be expected. Because the two currents are 120 degrees out of phase, a 15 percent loss of voltage occurs. The total voltage is actually only 208.

THE THREE-POLE CIRCUIT BREAKER

When the three phases are used individually as 120-volt lines, a single circuit breaker is placed in each hot line. In instances where the three phases are powering a three-phase load such as a motor, a three-pole circuit breaker can be used. The idea is the same, all three hot lines must have overcurrent protection (Fig. 12-17).

You cannot put a circuit breaker or fuse in one or two of the hot lines and hope that, if a fault occurs, it will occur between the fused lines and not in one of the lines you neglected to fuse.

A fault can occur between any one of the fused lines and ground. Or a fault can occur between any of the hot lines. Wherever the fault, if all of the hot lines are fused, then any excessive current will open one or more of the devices and prevent a hazard.

The pages of an electrical catalog show one-pole, two-pole and three-pole circuit breakers. You'll notice that the circuit breakers are rated in both amperage and voltage.

The amps range from 15 to 50 for one-pole breakers and 50 to 100 for two- and three-pole breakers. The two- and three-pole breakers are used in 240-volt lines and the one-pole breakers in 120-volt lines. To use the one-pole type in 240-volt lines, two breakers are installed, one for each hot line.

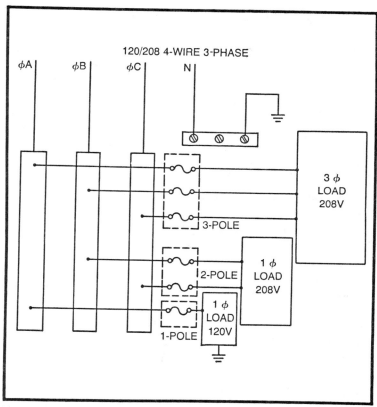

Fig. 12-17. Three φ loads require a 3-pole breaker or fuse system. Single φ loads can need 2-pole or single-pole breakers.

WIRING LARGE SWITCHES

A large switch is termed "a disconnecting means." This type of main switch is also wired in series with the hot lines. When the switch is opened, the hot lines are opened.

The switch and circuit breaker or fuse are located near each other. They are connected in series with each other. Both devices are required.

While the circuit breaker will open automatically during overload to protect the electrical installation, the main switch doesn't open during overload.

The main switch must be opened manually to kill the current. If you are working on a circuit, you can kill the current by opening the main switch. The switch can also be fitted with a lockout means so that you can lock the switch open and no one can accidentally close the switch and electrocute you.

In many residential panels, the main switch and main overcurrent device are combined in a plastic pullout box. The box has a handle on it. Within the box are two fuses of large size. When you pull out the box, you remove the fuses and the circuit is switched open. The circuit cannot be made live again until the pullout box is reinserted (Fig. 8-3).

The pullout box is a switch, overcurrent device and a form of lockout. No one can start the current flowing again if you have the pullout with you (unless they are carrying a spare pullout box, which is highly unlikely).

Section 230-71(a) of the Code prescribes that the service disconnecting means "...shall consist of not more than six switches or six circuit breakers mounted in a single enclosure, in a group of separate enclosures, or in a switch board."

The intent of this section is to insure that all electrical equipment can be disconnected with no more than six hand operations. In case of fire or explosion this can be a serious factor.

Some local ordinances do not permit even as few as six switches. They want a single main disconnect. Breaking of the circuit is faster this way.

The importance of this six-switch Code ruling will become clearer as you actually install electrical equipment. For example, if you can split up a main switch installation which is carrying current to six meters into six smaller switches and thinner conductors, the arrangement would be a lot less expensive and still perform with equal safety.

LOW VOLTAGE WIRING

Some appliances in the home only need between 6 and 10 volts to operate. A doorbell is an example. A chime needs between 15 and 20 volts. A switch activated by a pushbutton can operate on less than 30 volts.

The switch could turn lights and appliances on and off. The switch would be activated by remote control.

The above types of appliances and devices come under the category of low voltage wiring. Low voltage can be supplied in two ways.

One way is with a battery. A battery is DC and is available in various sizes. A door buzzer can be installed with a six-volt battery and not have any connection to the home wiring system whatsoever.

Another way to supply the voltages required for these low voltage circuits is with a step-down transformer.

The transformer takes 120 volts from the house supply into its primary. The voltage is then stepped down to about 8 volts in the secondary of the transformer. The 8 volts operate the device.

Such devices never need more than about 5 watts of power. An 8-volt device consuming 5 watts uses 1.6 amps (I = E/P 8/5 = 1.6). This small amount of electricity can easily be carried by even a poorly insulated #18 wire. This sort of wire is called bell wire and is available even in hardware stores.

The Code is not concerned with these low voltage circuits, except where they connect to or run near full-voltage circuits.

You can feel free to fish bell wire through walls, along baseboards, or wherever you like. There is no danger of fire or shock under ordinary circumstances.

The restrictions arise if you approach the full-voltage equipment. You are not permitted to run bell wire in the same pipe or armor with full-voltage wires. Low-voltage wires must throughout the run be kept at least two inches away from full-voltage wires. This includes keeping the bell wire away from boxes.

You are not permitted to install low-voltage wires in boxes, unless a foam or porcelain tube is installed to isolate the low-voltage wires, or, unless you install a metal barrier of the same thickness as the walls of the box to separate completely the low-voltage wires and the full-voltage wires.

DOORBELL TRANSFORMERS

The doorbell transformer is rated in voltage, cycles or wattage. The important consideration is the voltage (Fig. 12-18). For example, if you have to power a 10-volt doorbell, then you need a transformer that has a primary of 120 volts and secondary of 10 volts. Sixty cycles are the normal and most doorbell transformers can handle that. The wattage usually ranges between 5 and 10 watts. (It could be rated in VA, volt amperes, which in this case can be considered wattage.)

The transformer can be surface-mounted directly into a ½-inch knockout in a round or square outlet box, or on a switch or panel box.

There is about 6 inches of primary lead wire of at least size #14 which attaches to any 120-volt circuit and neutral.

The secondary can have terminal screws to which you can attach the bell wire. There is no need to ground the secondary.

Most transformers are small about 2 inches or 3 inches square.

HANDI-MOUNT TRANSFORMERS

PATENTED LOCK-ON AND CLEARANCE
FOR METAL BOX PLATE

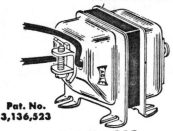

Insert LOCK-ON thru any ½" knockout. Tighten screw. Automatically locks, in any position. Or surface mount in regular manner.

Pat. No.
3,136,523

Cat. No. 285
FOR BELLS AND BUZZERS
PRIMARY—115V—50/60 Cycles
SECONDARY—10V-5VA

Cat. No. 286
FOR CHIMES
PRIMARY—115V—50/60 Cycles
SECONDARY—16V-10VA

SIZE: 2⅜" WIDE x 2⅝/₁₆" HIGH

Easy to install for both ½" knockouts and surface mounting. Can be mounted on round or square outlet box cover, on side of outlet box, switch or fuse cabinet. Insert patented "LOCK-ON" thru any ½" knockout and tighten screw. It automatically locks in any position. Six inch primary leads.

Secondary connects to 2 external terminal screws. Has low current consumption, low core loss and thermostatic temperature safety control. Fully insulated. Zinc Irridite finish. Wiring diagram and mounting instructions on each box.

INDIVIDUALLY BOXED

UNDERWRITERS' LISTED   **CSA App. No. 33564-1**

DOUBLE COIL BELL 2½"

OPEN GONG—
ENCLOSED TERMINALS
Cat. No. 292
USE 6-10V AC (TRANSFORMER)
OR 3-6V DC BATTERY

Non-adjustable. Clear, resonant gong. Easy to install. Has snap-on cover, high grade magnet wire coils, attractive grey finish. Operates on 3-6V DC or Eagle No. 285 or No. 287 10V Bell Transformer. Wiring diagram on each box.

INDIVIDUALLY BOXED

ENCLOSED TERMINALS
Cat. No. 296
USE 6-10V AC (TRANSFORMER)
OR 3-6V DC BATTERY

Non-adjustable. Gives clear, easy to hear signal. Has snap-on cover, high grade double magnet coils, attractive grey finish. Operates on 3-6V DC or Eagle No. 285 or No. 287 transformer. Wiring diagram on each box.

INDIVIDUALLY BOXED

Fig. 12-18. A bell transformer reduces the 120- or 115-volt input to about 10 volts to power the bell.

DOORBELL CIRCUITS

The typical doorbell circuit is an ordinary basic circuit consisting of the source, conductor, switch and load. The source is the secondary of the transformer, the conductors are the lengths of #18 bell wire, the switch is the pushbutton, and the load is the electromagnet in the bell itself (Fig. 12-19).

The pushbutton is a single-pole switch which is kept in the open position by a spring. When you push the button, you close the switch, allowing current to flow and causing the bell to ring.

The transformer can supply more than one bell. For instance, there can be a bell at the front door and a buzzer at the rear door. Both bell and buzzer circuits can be supplied with one transformer.

You must make sure that the combined wattage of the bell and buzzer does not exceed the wattage rating of the transformer. Suppose the bell uses 5 watts and the buzzer 5 watts. Then you

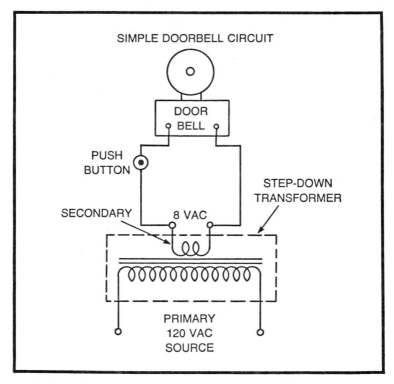

Fig. 12-19. The doorbell needs a source of 8 volts. The source can be AC or a DC battery.

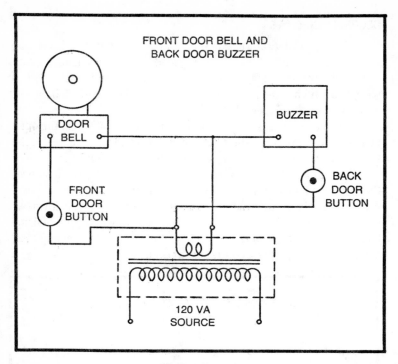

Fig. 12-20. Doorbells, buzzers and chimes can be wired in a number of different ways. Here is one way.

cannot use a transformer with a rating of only 5 watts. Most of the time you could get away with it since it is unlikely that the front door and back door pushbuttons would ever chance to be actuated simultaneously. Yet it could happen.

The price difference is negligible, so use a 10-watt transformer.

The two circuits are wired in parallel as the schematic shows (Fig. 12-20). They are both ready to ring or buzz at all times.

When you have just one bell and you need a number of pushbutton locations, then you wire the circuit as a basic circuit for the first button. The rest of the buttons are all wired in parallel, so that whichever button is pushed, the circuit closes and the bell rings (Fig. 12-21).

Some installations require one pushbutton and three bells. The three bells are wired in parallel and the pushbutton in series with all three of them (Fig. 12-22). (If you want to add more pushbuttons, wire them in parallel with the original button.)

When the pushbutton is pressed, all three bells receive current and all three bells ring.

You can add as many bells in parallel as you want and obtain the same effect. Just be sure that the transformer can carry the load of all the bells.

LOW VOLTAGE LIGHTING CONTROL

If you think about the wiring in the three-way and four-way switch lighting circuits, you can see it can become very cumbersome. Even if #14 wire is used, much copper wire will be used and a number of connections will have to be made. The wires run back and forth and get complex. The switches have 3 and 4 connections each. In a long hallway hundreds of feet of #14 wire would be used. Some localities insist on #12 wire, which is even more cumbersome and expensive.

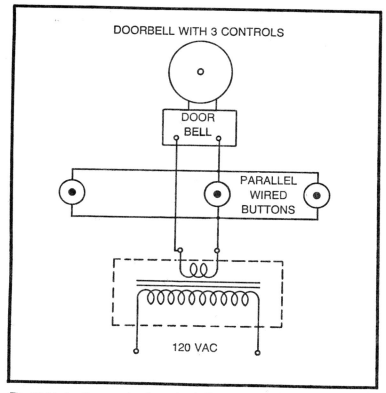

Fig. 12-21. Another way to wire a doorbell.

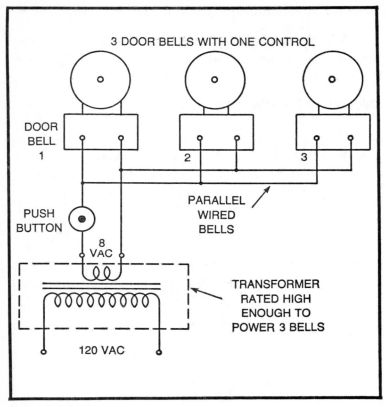

Fig. 12-22. A third way to wire a doorbell.

There is an easy way out. If you use low-voltage wiring, you can substitute #18 bell wire for much of the #14 or #12 wire. The bell wire is easy to handle, is governed by few or no Code safety restrictions, and is relatively inexpensive.

Instead of the three-way and four-way toggle switches, push-buttons can be used. These are the advantages of remote-controlled, low-voltage switching systems.

The system is based upon a relay which turns on and off. The relay is installed in the lighting fixture box. The relay can switch the light on and off (Fig. 12-23).

The relay has two ends. One end wires into the light and the 120-volt circuit. This end of the relay acts like a toggle switch. It is wired in series with the light and, as the relay closes, the light goes on. It is rated at 15 amps or 20 amps.

The other end of the relay is not connected electrically to the 120-volt circuit. It is attached to the low-voltage circuit. The lighting circuits are usually about 24 VAC. When the 24 VAC flows into the relay, an electromagnet turns on and moves the relay arm. This physically switches the 120-volt circuit on or off.

You can install as many pushbuttons in the 24-volt circuit in parallel as you please. Pressing any one of them activates the relay. If the relay is in an OFF position, the light will go on. If the relay is in an ON position, the light will go off.

WIRING THE LOW VOLTAGE LIGHTING

Three wires emerge from the low-voltage end of the relay. The three wires come from two coils. One is called the ON coil and the other is the OFF coil (Fig. 12-24).

The schematic shows that each coil has its own pushbutton. There is an ON button and an OFF button. When the ON button is pushed, the ON coil receives current and is energized. As an electromagnet, it attracts an armature as shown. The armature is at-

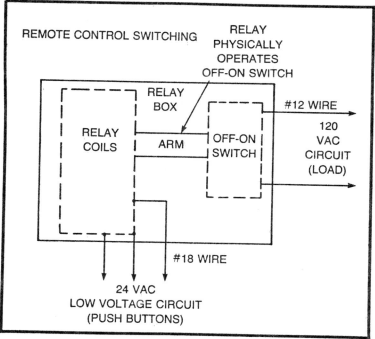

Fig. 12-23. Remote control light switching can save a lot of copper and conduit.

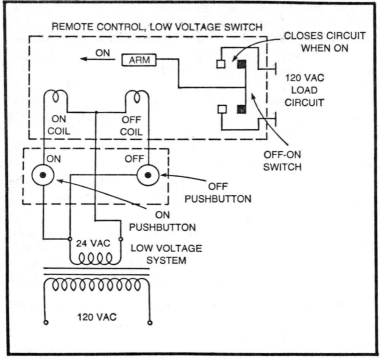

Fig. 12-24. The remote control switch moves an armature between the low-voltage and full-voltage systems.

tached to the switch in the relay box. The movement of the arm turns the switch on.

To turn the switch off the OFF button is pressed. The armature is attracted back to itself. The moving arm disconnects the switch and physically turns the switch to OFF.

The three wires from the two coils are coming from the two ends of the coils and the center tap. The two coils form two circuits in parallel. Each circuit is pushbutton-controlled.

While the two pushbuttons are spring loaded to stay in the OFF position, except when held down, the armature is not spring loaded. It stays in the OFF or ON mode until it is moved. A momentary push on a button is all that is required to energize a coil and move the armature.

The beauty of the scheme is that the relay switch handles all the heavy lighting current. The low-voltage system handles practically no current, yet controls all that heavy current.

RESIDENTIAL WIRING

Earlier in this book we touched upon the actual laying out of a residential wiring job. At some time or other you may be confronted with the task of designing a system in which the types and number of loads are previously specified. You'll draw a rough sketch to work out the load problem (Fig. 13-1).

LAYING OUT A JOB

There is 1700 square feet of house, 300 square feet of open porch and 500 square feet of garage. It is a nice-sized home. The electrical service is three-wire, 120/240V.

The appliances and the size of their respective loads are given:

> 2.5 kW Dishwasher
> 1.0 kW Trash Compactor
> 1.0 kW Garbage Disposer
> 15.0 kW Range
> 5.0 kW Water Heater
> 12.0 kW Air Conditioner
> 15.0 kW Heating Strip System
> 6.0 kW Dryer

The first question that comes to your mind is what is the total load. Once you know the total load, then you can calculate the service requirements.

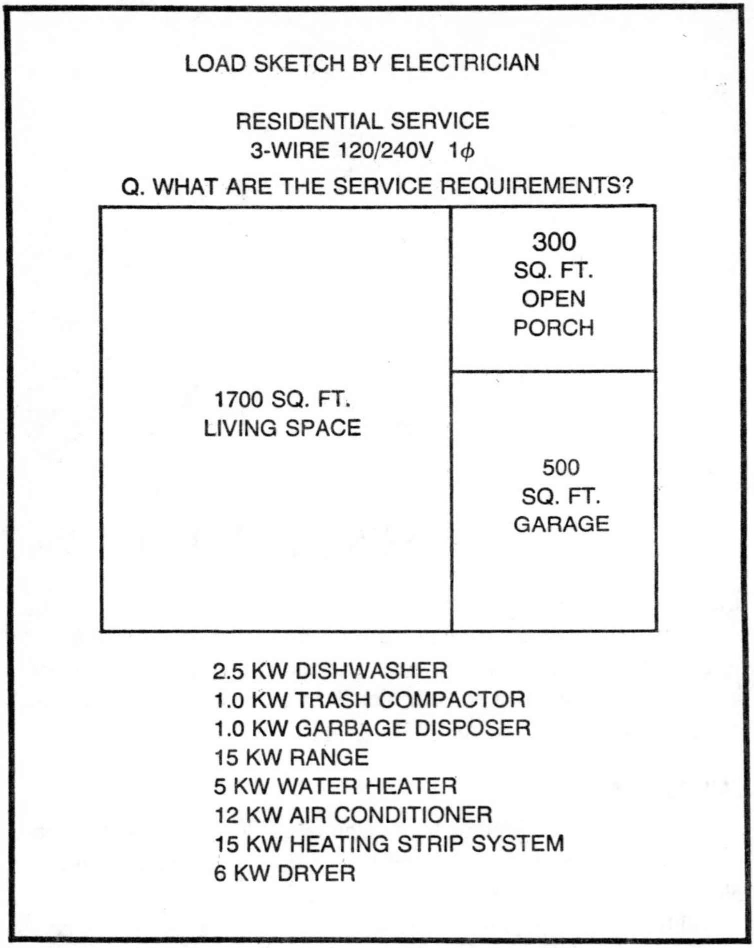

LOAD SKETCH BY ELECTRICIAN

RESIDENTIAL SERVICE
3-WIRE 120/240V 1φ
Q. WHAT ARE THE SERVICE REQUIREMENTS?

1700 SQ. FT. LIVING SPACE	300 SQ. FT. OPEN PORCH
	500 SQ. FT. GARAGE

2.5 KW DISHWASHER
1.0 KW TRASH COMPACTOR
1.0 KW GARBAGE DISPOSER
15 KW RANGE
5 KW WATER HEATER
12 KW AIR CONDITIONER
15 KW HEATING STRIP SYSTEM
6 KW DRYER

Fig. 13-1. An electrician can make a rough sketch of the load that he is expected to install wiring for.

First of all, you must determine the size of the three service conductors, that is, the two phase lines and the neutral wire. Once the conductor sizes are determined, the conduit size can be calculated, the size of the main switch and panel becomes known, and the grounding conductor size can be located in the Code book chart.

FIRST GLANCE

An experienced electrician can glance at these load requirements and quickly conclude that a 20-amp service will fill the bill and even provide a goodly amount of extra capacity in the form of spare circuit breaker space in the panel for the future.

If the homeowner will accept the expense, that is the system which will be installed. How did the electrician arrive at this figure? Through his years of wiring experience with homes of this size.

This procedure, whereby experience gives the electrician the answers, is all well and good. However, it doesn't always work. Not all homeowners will consent to the spare-no-expense type of electrical installation, especially if a competitive electrician points out the fact that a smaller service would save a lot of money, be completely satisfactory, and yet meet Code requirements (Fig. 13-2).

An electrician can lose a job or have an unhappy customer on his hands after an unnecessarily expensive job is installed. According to the Code, load calculations can be demanded by large percentages, yet provide a safe and satisfactory installation at a reasonable cost. The Code's attitude is that there is no need for a lot of spare circuit breaker spaces in a panel, or for the extra expense of empty space.

You should be able to calculate the minimum service requirements according to Code. Perhaps you'll use service conductors a

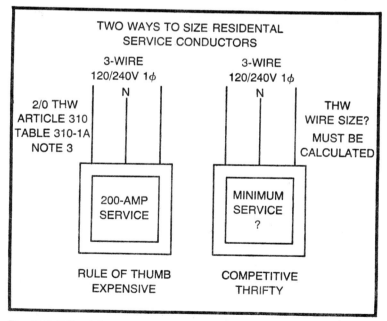

Fig. 13-2. If the dollars are available, this given load can receive a 200-amp service. However, a carefully calculated minimum service is usually thriftier.

size or two larger when the dollars are there, but you should be able to bid on a job and get it when the extra dollars are not there.

Therefore, the total load becomes the total demanded load. From the total demanded load you can calculate the three service conductors. Once you know their size, everything else falls into place.

BRANCH CIRCUITS IN GENERAL

After the size of the service conductors has been determined, you must calculate the size of the branch circuit conductors. Even though you demand the service conductor capacities by large percentages, you demand the branch circuits hardly at all.

When a branch circuit is conducting, it could often need all the ampacity it is rated to receive. When electricity is being fed to the elements of the water heater, the conductor must be able to stand the load. When the garbage disposer is grinding away, the full wattage rating is being used.

The Code in Article 210-23 (a) , (b), and (c) tells you how to rate the branch conductors. The 15- and 20-amp circuits must have conductors of certain sizes. If a stationary appliance is energized, the ampacity of the wire must be larger than the appliance wattage. The appliance must not pull any more current than 80 percent of the branch ampacity.

For example, a small clothes dryer is a stationary appliance. If it is a 240-volt appliance and is rated at 3 kW, then it draws

$$I = P/E = 3 \text{ kW}/1240 = 12.5 \text{ amps}$$

Eighty percent of a 20-amp circuit is 16 amps. The dryer can be used on that kind of circuit easily. Other examples of stationary appliances are self-contained ranges, window air conditioners and similar items which are normally left in one location but can be moved if necessary because they are hooked up by cord and plug.

When a fixed appliance is energized, the 15- or 20-amp branch circuit shall not exceed 50 percent. Fixed appliances are water heaters, oil-burner motors, room air conditioners, built-in ovens, cooktops, garbage disposers, trash compactors and the like.

The 25-, 30-, 40- and 50-ampere branch circuits are not required to carry extra ampacity for fixed appliances. If a cooktop oven draws 50 amps, you can install a 50-amp branch circuit. The cooktop doesn't need the overrating.

In 25- and 30-amp circuits, stationary appliances do need the extra ampacity though. Study Article 210-23 of the Code. The important thing is that you cannot attach a load that exceeds the branch ampacity. This is not like the service entrance where you demand large amounts of ampacity.

ARTICLE 220

Getting back to the service entrance conductors, Article 220 covers the procedures you must follow if you want to demand the installation. Since all of the electricity that enters the house must pass through the service entrance conductors, these are calculated first. The service entrance conductors are considered to be "feeders." From here on we'll refer to them as feeders (Fig. 13-3).

To calculate the feeder ampacity, each and every branch circuit in the residence is examined carefully and its feeder ampacity demanded, if allowed by the Code. This is done by computing the kilowattage consumption of each branch circuit. Ampacity and kilowattage are at either end of the Ohm's Law formula. If you know one, you can quickly calculate the other.

For example, 10.8 amps at 240 volts is $10.8 \times .240 = 2.6 \, kW$. Or 4.3 kW at 240 volts is $4.3 \, kW/.240 = 18$ amps. (The .240 is used

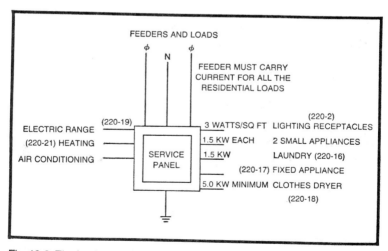

Fig. 13-3. The feeders must have enough ampacity to carry all of the residential load currents.

to automatically convert watts to kilowatts. This is a common practice of electricians.)

The way to calculate total feeder ampacity is to add up all the demanded kilowattages and convert them to amperes.

Article 220 covers dwellings load by load from 220-1 through 220-30. Later sections of Article 220 deal with larger buildings and installations. These topics will be discussed in Chapters 14, 16, and 17.

MAXIMUM DEMAND AND MINIMUM WIRE SIZE

If you use the demands in the Code to the limit, you refer to Article 220-30 (Optional Calculation). While the Code does permit extensive demanding in 220-30, some local jurisdictions do not. You can get advice from the local electrical inspectors as to what they will and will not allow.

Those of you readers who are striving for Journeyman status will find that most examinations will require you to calculate a dwelling according to Code, as this chapter shows. Learn these minimum wire size calculations and then, if need be, adjust them to conform with the requirements of the locality you work in.

The important parts of Article 220 are Sections 2, 16, 17, 18 and 19. Reference is also made to Sections 21 and 22.

Article 220-2, especially (b) and Table 220-2 (b), refers to lighting and receptacle loads. Article 220-16 deals with small appliance and laundry loads. Article 220-17 applies to fixed appliances and 220-18 applies to a stationary appliance, the electric clothes dryer. Article 220-19 explores the complexities of electric ranges. Article 220-21 is largely devoted to heating and air conditioning. Article 220-22 helps you calculate the size of the neutral wires.

LIGHTING AND RECEPTACLE LOADS

The lighting in the home and many of the 15-amp and 20-amp receptacles are calculated by means of the square footage of the home.

The Code in Article 220-2 (b) is very specific about the lighting and receptacle load. In Table 220-2 (b) under Dwellings, the Code requires 3 watts per square foot of floor space. Our example dwelling has 1700 square feet, plus 300 square feet of open porch and 500 square feet of garage.

The Code says that the computed floor area shall not include open porches and garages. This means that the wattage is computed according to the living area only. There is 1700 square feet of living area. 1700 sq. ft. × 3 watts/sq. ft. = 5100 watts or 5.1 kilowatts. (Record this figure for your eventual total.) Therefore, we have a total lighting and receptacle load of 5.1 kilowatts. There is no demanding of this individual load. The minimum is already 3 watts per square foot. Many dwellings are built with 4, 5 or more watts per square foot, but 3 is the minimum.

The '71 Code recommended that there be a lighting-receptacle circuit for every 500 square feet. This would compel our example dwelling to have between 3 and 4 of these circuits. Actually, while the minimum is safe, more circuits would offer greater convenience.

If the circuits are wired with #14 wire, then they are 15-amp circuits. Each 15-amp circuit at 120 volts (15A × 120V) would yield 1800 watts or 1.8 kilowatts.

There are 2.8 of these circuits in a 5.1-kilowatt total load. It would be best to run 5 circuits, if you can get paid for it. With the extra capacity you could run outlets to a basement or unfinished attic or to other unfinished spaces so that the homeowner would be ready when future needs present themselves.

SMALL APPLIANCE LOADS

The general light-receptacle circuits just discussed can all be of the 15-amp, #14 wire type. It is recommended though that some or all of them be of the 20-amp, #12 wire type. Some local jurisdictions dictate that #12 wire be used. However, it is less expensive to run #14 if it is allowed.

These 15-amp circuits can be hard pressed if they have to power food heaters and warmers and other modern day heavy-drawing, plug-in appliances. Many of these appliances consume 1000, 1200 and even 1800 watts.

As a result the Code requires, in addition to the 3 watts per square foot receptacles, two appliance receptacle circuits. These circuits run to the kitchen, pantry, dining room and family room.

Article 220-16(a) of the Code prescribes that each of these circuits be wired only with #12 wire and be fitted with a 20-amp breaker or fuse. Each circuit will have an ampacity (20A × 120V) of 2400 watts or 2.4 kilowatts. In your load calculation, you count them as 1500 watts each. The 2400-watts capacity is present, but you'll hardly ever be using more than 1500 watts on each one (a total of 3000 watts) at the same time.

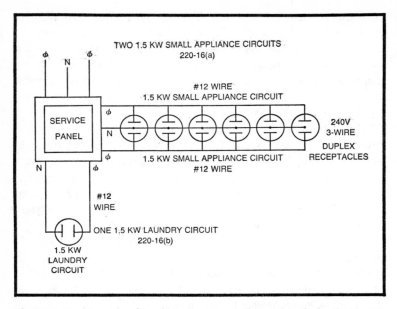

Fig. 13-4. The two sides of each duplex receptacle can be attached to the two sides of the 3-wire circuit with a common neutral.

The best way to install the two circuits is to install both of them on a single three-wire, 240-volt circuit. The two sides of each duplex receptacle can be attached to the two sides of the three-wire circuit with a common neutral (Fig. 13-4).

In that way the two circuits balance each other. Also, if two large appliances are plugged into the same duplex receptacle, they are really plugged into two different circuits.

Other advantages of using one three-wire circuit in place of a pair of two-wire circuits are less voltage drop and elimination of one length of conductor wire.

LAUNDRY CIRCUIT LOADS

A companion to the small appliance circuits is the required laundry circuit, when there is a laundry in the dwelling. It does not have to be a laundry room as such; it need only be an area where the washing of clothes occurs. A washer and dryer are present (Fig. 13-4).

Articles 220-16 (b) and 220-3 (c) of the Code discuss the laundry. There is to be a laundry receptacle. It must be on an individual circuit. The circuit must be a 20-amp circuit, and that means #12 wire.

In the load calculation you add another 1500 watts or 1.5 kilowatts.

CONVENTIONAL CALCULATIONS

Earlier in this chapter, it was mentioned that Article 220-30 (Optional Calculation) of the Code will provide the safe minimum wire size. This is true, but some local jurisdictions will not permit you to use optional calculations. Also, if the total ampacity is under 100 amps, you cannot use optional calculations. There is a conventional way to calculate the load. The conventional way has some demanding to it, but not nearly the amount of demanding of the optional method.

At this point let us pause and calculate the total load in the conventional way (also known as the long way), even though this residence is well over 100 amps.

So far there are 5.1 kilowatts for the square footage, 3.0 kilowatts for the small appliance circuits, and 1.5 kilowatts for the laundry circuit. 5.1 kW + 3.0 kW + 1.5 kW = 9.6 kW. Article 220-11 (the general lighting load) of the Code has a table that gives you demand factors.

The first 3.0 kilowatts are taken at 100 percent. The remainder, up to 120 kilowatts, can be demanded at 35 percent.

```
1st 3.0 kW at 100% =
Remainder 9.6 − 3.0 = 6.6 kW                    3.0 kW
                    6.6 kW at 35% =             2.3 kW


      Total Lighting-Receptacle =               5.3 kW
```

The general demanding allowed in the conventional calculations applies to the range, the heater and the air conditioner. The range still gets special demanding. The 15-kilowatt range demands out at 9.2 kilowatts.

You do not add the heater *and* the air conditioner, just the largest one. The largest load is the air conditioner at 12 kilowatts. (The heater is a strip and gets special demanding from 15 kilowatts to 9.75 kilowatts, which is smaller than the wattage of the air conditioner.)

Adding up all the loads,

Lighting-Receptacles	-	5.3 kW
Dishwasher	-	2.5 kW
Garbage Disposer	-	1.0 kW
Trash Compactor	-	1.0 kW
Water Heater	-	5.0 kW
Dryer	-	6.0 kW
Range	-	9.2 kW
Air Conditioner	-	12.0 kW
Total Load	-	42.0 kW

the service needed for 42.0 kilowatts (42,000 watts) is

$$\frac{42,000W}{240V} = 175A$$

Looking up 175-amp service in Table 310-16, Note 3, shows that you need #1 copper conductors. All of the other requirements can now also be looked up as easily. It might be interesting to notice and remember as a rule of thumb that a bit more than 4 amps are needed to produce a kilowatt with 240-volt electricity.

GETTING BACK TO OPTIONAL

It is somewhat simpler to calculate the service requirements by the conventional method since there is less demanding. As you'll see, though, the actual installation can be a lot more expensive.

When the requirements are under 100 amps, there is no choice, but when there is a choice, a lot of money can be saved; especially if it's a housing development where the savings are multiplied by the number of homes built.

Table 220-11 of the Code book is used only for the conventional calculation, not for the optional. For the optional you use Table 220-30, which we will get to shortly.

FIXED APPLIANCE LOAD

Every home has a number of fixed appliances. Our example home has a normal complement: dishwasher, trash compactor, garbage disposer, water heater, air conditioner, electric heater, clothes dryer, and possibly the range.

Article 220-17 of the Code gives you a demand factor of 75 percent of the nameplate rating load of the fixed appliances. There must be a minimum of four appliances; otherwise you cannot demand them at all.

At first it appears as if there are far more than four, but hold it! An exception is printed in italics at the bottom of the page: exclude ranges, dryers, heaters and air conditioning.

Fortunately, the dishwasher, compactor, garbage disposer and water heater still remain. These four loads can be added together and then demanded by 75 percent.

Dishwasher		2.5 kW
Trash Compactor		1.0 kW
Garbage Disposer		1.0 kW
Water Heater		5.0 kW
	Total	9.5 kW

9.5 kW at 75% = 7.1 kW. That is the total ampacity of the four fixed appliances needed in the feeder. Remember, the branch circuits still need the full nameplate ampacity required for each fixed appliance. It is the feeder capacity that is being demanded.

If the trash compactor had not been included in the home, the demand would have been raised to 100 percent, since there would have been only three fixed appliances.

CLOTHES DRYER

The load for the clothes dryer is calculated according to the nameplate value, but values of less than 5 kilowatts must be raised to the 5-kilowatt minimum for calculating demand.

When the nameplate of the dryer reads more than 5 kilowatts, then the actual nameplate rating is to be used.

There are occasions when a dryer is to be purchased at a later date and the nameplate rating is not available. In these cases, you are permitted to install a 5-kilowatt capacity in the feeder for the dryer. It is tacit that a dryer no larger than 5-kilowatt is permitted. To install a larger than 5-kilowatt dryer in a circuit calculated to accommodate a 5-kilowatt dryer would be in violation of the Code.

The Code book discusses clothes dryers in Article 220-18.

ELECTRIC RANGES

The electric range in this example is a 15-kilowatt model. This constitutes a heavy load. If every burner and the oven were operating at the same time, the range would pull 62.5 amps (15 kW/.240 = 62.5A) through the feeder.

Chances of all heating components being on at the same time are slim. Therefore, the range in a home can get good demand percentages.

For example, if a range has a nameplate rating of 12 kilowatts, you need compute only 8 kilowatts for the range feeder ampacity.

The Code provides a demand chart (Table 220-19) for the calculation. The complete chart applies to all buildings (including apartment houses), commercial electric cooking, etc. More than four ranges are found in large buildings, but rarely in single dwellings.

Even if you have two separate wall ovens and a cooktop, they need not be computed separately for the feeder calculation. Add all of the wattages together and compute the total of the range loads as if the residence had but a single range.

The calculations are spelled out in Notes 1, 2, 3, and 4 of Code book Article 220-19. These notes and the Article's table are complex and must be studied carefully before they are applied. There is a quick way though to arrive at a total range load figure for a single range.

Impress your mind with the fact that a 12-kilowatt range is demanded to 8 kilowatts. Then for each kilowatt (or fraction of a kilowatt) above 12 kilowatts, add 400 watts to the 12 kilowatts.

For example, a 13-kilowatt range would be demanded to 8 kilowatts + 400 watts. That's 8.4 kilowatts. A 14-kilowatt range is 8 kilowatts + 800 watts. That is 8.8 kilowatts. A 15-kilowatt range is demanded by 8 kilowatts + 1200 watts, which equals 9.2 kilowatts.

If it is a 15.6-kilowatt range, then the extra fraction is computed as 400 watts more. That would cause 1600 watts to be added to the 8 kilowatts. The 15.6-kilowatt range therefore demands out to 9.6 kilowatts.

ANALYZING EACH LOAD

So far in this chapter we have examined the loads that are called "other loads" in Table 220-30 of the Code book. They add up to the following in our 1700-square foot example dwelling.

CODE ARTICLE	LOAD		KW IN φ LINES	KW IN N	
220-2	Sq. Footage		5.1	5.1	
220-16	2 Small Appliance Cir.		3.0	3.0	
220-16	Laundry		1.5	1.5	
220-17	(Dishwasher	2.5)	7.1		
	(Trash Com- pactor	1.0)		1.0	
	(Garbage Disp.	1.0)		1.0	
	(Wtr. Htr.	5.0)			
	Total	9.5			
220-18	Clothes Dryer		6.0	4.2	70%
220-18	15-kW Range		9.2	6.4	70%
	"Other Load"		Total 31.9	22.2	

In the two phase lines of the 120/240V, three-wire service, each phase line must be able to carry its eventual demanded load. Up to this stage in the calculations, if all of the lines are operating, they will consume 31.9 kilowatts.

What about the single neutral wire? How much load will it carry while these two phase lines are in full operation?

Looking back at the chart we've just read, examine each load at a time. The first six loads have the same neutral current as the phase. The seventh load, the water heater, has no neutral wire, so no neutral feeder ampacity is needed.

The clothes dryer and range both have neutral wires. If you use the same size neutral as the phase line, you will pass inspection, but you will install costly extra ampacity that will never be used.

To avoid the waste of copper, the Code allows you to demand the neutral feeder range load by 70 percent. Since the clothes dryer uses an almost identical circuit, it can also be demanded by 70 percent. This makes the neutral feeder load for the dryer 4.2 kilowatts and for the range 6.4 kilowatts.

The total neutral feeder capacity for the "other load" totals out to 24.7 kilowatts.

USING TABLE 220-30

Okay, we now have the "other load." You must have noticed that neither the heater nor the air conditioner is one of the parts of the "other load." They are treated separately in Table 220-30 of the Code book.

The best way to insert them into your computations is to list them as they stand.

CODE ARTICLE	LOAD	KW ϕ	N	DEMAND FACTOR
220-21	Air Conditioner	12.0	-	100%
220-30(5)	Heating Strip (15 kW × 65% = 9.75) Smaller Than A/C	-	-	65%
TABLE 220-30	1st 10-kilowatt Other Load	10.0	10.0	100%
	Remainder Of Other Load (ϕ 21.9, N 14.7 × 40%)	8.8	5.9	40%
	Total Demanded Feeder Load	30.8	15.9	

The air conditioner is not demanded. There is a 12-kilowatt load. When the air conditioner is on, it is on all the way. Because there is no neutral wire in the air conditioner, the feeder neutral doesn't have to bear any current as far as the air conditioner goes.

The heating strip comes under another category. You are permitted to demand the electric heating by 65 percent. It does not have a neutral, either.

Then you can apply Code book Article 220-21 (Non Coincident Loads): since the two units will never be on together, the smaller load can be crossed out. That is the heater. Count only the 12-kilowatt air conditioner.

Then look at the "other load." Mark the first 10-kilowatt load at 100 percent. The remainder of the phase load $(31.9 - 10.0 = 21.9)$ is marked in. Then the remainder of the neutral $(24.7 - 10.0 = 14.7)$ is added.

The remainder is then demanded at 40 percent. After everything is added together, the phase lines must each carry 30.8 kilowatts and the single neutral line must carry 15.9 kilowatts.

Converting to amps, the phase lines and neutral line have to be thick enough to carry 128 and 66 amps, respectively.

WIRE SIZES

Once the ampacity of the feeder load has been calculated, you can lay out the rest of the service needs. It is all in the Code book. Just look it up.

For the residential feeder wire size, you refer to the Notes of Table 310-16. In Note 3 the wire size for the phase lines is given. The residential system we are laying out needs 128 amps.

The next largest size is 150-amp service. The Note specifies the use of #1 copper in RH, RHH, RHW, THW or XHHW. If you want to use aluminum or copper clad with these insulations, use #2/0.

The neutral wire would pass inspection if it were of the same size as the phase lines. However, the minimum neutral can be obtained from Table 310-16. The neutral feeder load is only 66 amps. In these insulations 66 amps needs a #4 copper or a #3 aluminum.

The grounding conductor, the wire from the panel to the actual ground, is obtained from Table 250-94.

The largest conductors are the phase lines of #1 wire. For a #1 copper phase line, a #6 copper grounding conductor is allowed. However, use the neutral size #4. If you use the #2/0 aluminum as the phase lines, then you'll need a #2 aluminum grounding conductor.

The panel is 150 amps, since it is the service switch and main overcurrent protection.

THE PIPE

If a conduit is used to carry the feeder, it is calculated in the following manner. It will house three conductors: two #1 coppers and one #4 copper. However, if aluminum or copper clad conductors are used, there will be two #2/0's and one #3. What are the pipe sizes?

	#2/0	.2781	sq. in.
Aluminum	#2/0	.2781	sq. in.
	#2	<u>.1473</u>	sq. in.
		.7035	sq. in.

Looking in Table 4, the next largest pipe is a 1½-inch size with 0.82 sq. in. This is the pipe that the aluminum is carried in (more than two conductors, 40 percent).

The next step, now that you know what you need in service equipment, is to figure out the branch circuits. That begins with balancing.

BALANCING THE BRANCH CIRCUITS

Now that we know what size service equipment is needed to carry the current in our example home, the next step is to decide where to place the branch circuits to obtain best efficiency.

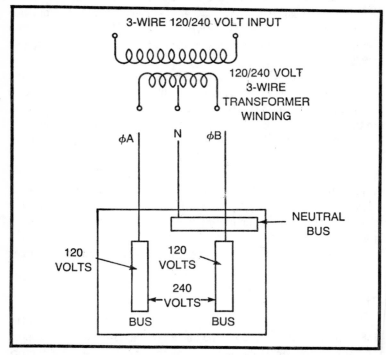

Fig. 13-5. In the 3-wire, 120/240-volt input, the phase is developed across one transformer secondary.

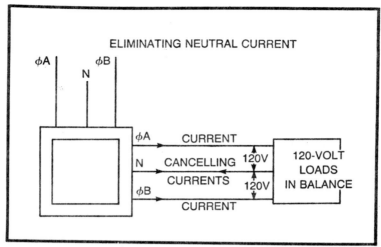

Fig. 13-6. The trick to planning a well balanced installation is eliminating the neutral current.

The key to planning the well balanced installation is the neutral current. The trick is to come as close to completely eliminating the neutral current as possible.

In the three-wire, 120/240-volt input, there is one phase (Fig. 13-5). The phase is developed across one transformer secondary winding. There are 240 volts across the winding.

The phase is 60-cycle and when one end of the winding is at 240 volts, the other end is at zero volts. The two ends keep reversing this voltage 60 times a second.

Therefore, as the current flows in one direction in one hot line, the current flows in the other direction in the opposite hot line. Even though there is only one phase, each line has a different direction of current movement, since the lines are at opposite ends of the secondary.

To obtain 120 volts from this sytem, the center of the winding is tapped to ground. This, of course, is neutral.

As the current in one hot line is returned to neutral in one direction, the current in the other hot line is sending neutral current in the opposite direction. If both currents are equal, no current flows in the neutral line (Fig. 13-6).

If the neutral line has only a negligible amount of current, it won't heat up and waste energy. It won't have any voltage drop and maximum efficiency will have been attained.

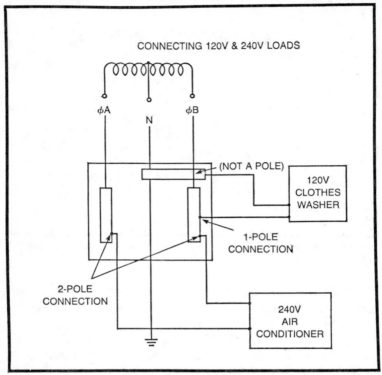

Fig. 13-7. A 120-volt load connects between a phase line and neutral. A 240-volt load connects between the two phase lines.

Balancing the branch circuits between the two hot lines accomplishes all of these desirable conditions.

BALANCING RULES

The balancing scheme for a three-wire, single-phase system is not difficult. Once you have your branch circuit kilowatts, you can do it.

The idea as stated is to determine where the loads go across the secondary winding. There are two ways the loads can be installed: the 240-volt loads across the entire winding, and the 120-volt loads across one end of the winding and the neutral center tap (Fig. 13-7).

When you balance, forget about demand factors. All wattages are computed as nameplate or maximum. The balancing is done with the wattage numbers. It can also be done with amps, but wattages are usually used.

Draw a chart like the illustration (Fig. 13-8). The two sides of the 120/240-volt, three-wire can be called A and B. Begin the balancing with the 240-volt loads. These are the loads that need two poles on a circuit breaker, or two fuses, each in a hot line.

These 240-volt loads are automatically balanced, since they are across the entire input. These are the air conditioner, heating system, and water heater. These loads have no neutral wire, just two phase lines that attach across 240 volts.

The 120/240-volt loads are also automatically balanced. These are the clothes dryer and range. They have a neutral. The neutral allows the appliance to have 120 volts to power lights, timers or other auxiliaries.

As a result, there is a small amount of neutral current, even though the appliance designer tries to provide as little neutral current as possible. If there are two such tiny 120-volt loads in an appliance, the designer will balance them by placing one across A and the other across B in the three-wire scheme.

THE 120-VOLT LOADS

Besides the 240-volt loads, there are 120-volt loads. In our example residence, there are two small appliance circuits, a laundry circuit, a dishwasher, trash compactor, garbage disposer and all the lighting and receptacle circuits. There are between three and six of these.

The two small appliance circuits are easy to take care of. Each is 1.5 kilowatts and one can be placed across A while the other one is placed across B.

The laundry circuit has 1.5 kilowatts and can be placed across A. The dishwasher has 2.5 kilowatts and is placed across B. The trash compactor has 1.0 kilowatt and is placed across A. The garbage disposer has 1.0 kilowatt and is placed across B.

If you add up all the loads so far, you'll find that B has 1.0 kilowatt more than A. What can be done?

Simply adjust the last load so balancing takes place. The 5.1 kilowatts that are going to be installed as lighting and receptacle loads can be split easily so that 3.1 kilowatts are placed across A and 2.0 kilowatts across B.

The final balancing is of 60.1 kilowatts across A and 60.0 kilowatts across B. This is a 100-watt difference, which is negligible.

Actually, all the loads in the dwelling will never be operating at the same time. This 100-watt difference will never actually happen.

TYPICAL ONE-PHASE BALANCING KW		
NO NEUTRAL LOADS 2-POLE	φA	φB
AIR CONDITIONER	12.0	12.0
HEATING SYSTEM	15.0	15.0
WATER HEATER	5.0	5.0
NEUTRAL LOADS 2-POLE		
RANGE	15.0	15.0
CLOTHES DRYER	6.0	6.0
120-VOLT LOADS		
SMALL APPLIANCE	1.5	1.5
LAUNDRY	1.5	
DISH WASHER		2.5
TRASH COMPACTOR	1.0	
GARBAGE DISPOSER		1.0
LIGHTING & RECEPTACLE	3.1	2.0
TOTALS	60.1 KW	60.0 KW

Fig. 13-8. A single-phase balancing chart considers the loads as 2-pole no neutral, 2-pole neutral, and 120-volt types.

The electrical system will rarely be in a position where a lot of neutral current is flowing in unbalance.

The neutral feeder of our previous calculations is able to handle 15.9 kilowatts with ease. The neutral feeder is a Code-permitted #6 wire. The optional calculation we computed permitted this #6 wire. (We used a #4 for added safety.)

When local codes will not allow such large demanding, then the neutral must be larger. However, that calculated 15.9 kilowatts (translated to 66 amps) will never travel in the neutral as unbalanced current if you balance according to the rules.

MAXIMUM UNBALANCE

The Code book refers to maximum unbalance in Article 220-22: "The feeder neutral load shall be the maximum unbalance of the load...."

Even though you balance the load to satisfy a condition under which every circuit in the house would be operating and there would be practically no feeder neutral current, you still must size the neutral for maximum unbalance.

What is maximum unbalance? A maximum unbalance occurs when all of the 120-volt circuits on one side are operating and none of the 120-volt circuits on the other side are operating. All of the neutral current possible runs through the neutral feeder.

For example, in our sample dwelling, what is the maximum unbalance? We have 60.1 kilowatts in A and 60.0 kilowatts in B. However, the 240-volt circuits won't work unless they are balanced, so you do not count them.

The 120/240-volt circuits of the dryer and range are slightly unbalanced because of the subordinate 120-volt circuits they contain. The Code book (in Article 220-22) considers the maximum unbalance range load to be 70 percent of the demanded load.

That means 70 percent of the 9.2-kilowatt demanded load of the range, not the 15-kilowatt nameplate rating. This makes the range and dryer have unbalanced loads of 6.4 and 4.2 kilowatts, respectively.

Add to those two loads the 5.6 kilowatts which are the total connected load between A and ground and which are greater than the total connected load between B and ground.

Add the three loads together and you'll get the maximum unbalanced load.

Range	6.4 kW
Dryer	4.2 kW
A to Ground	5.6 kW
Maximum Unbalance	16.2 kW

The 16.2 kilowatts convert to 67.5 amps (16.2/.240).

That is the maximum unbalanced Feeder Neutral Load. You can size the feeder to carry 67.5 amps.

For comparison, what was the feeder neutral size in the optional calculations? Computations performed according to Code book Table 220-30 yielded 15.9 kilowatts. That is 66 amps.

The 67.5 amps and 66 amps are essentially equal. Both use a #4 feeder neutral in THW copper. The maximum unbalance turns out to be very much like the unbalance you derive when you use the optional calculation of Code book Article 220-30.

SIZING THE BRANCH CIRCUITS

Sizing of branch circuit conductors can be done by electricians almost by reflex. Most residential wiring is #12, with an ampacity of 20 amps.

Many circuits would actually need only #14 wire since they'll be handling no more than 15 amps, but it is advisable to never use anything thinner than #12, even on a 15-amp circuit. The difference in price between #14 and #12 is minimal and it's worthwhile to have the extra ampacity for the future.

The 30-amp circuits use #10 wire, 40-amp circuits #8, and 50-amp circuits #6. Code Table 210-24 lists the ampacity of the various copper wires. It also lists the necessary overcurrent protection, which is the same as the circuit rating. A #14, 15-amp wire gets a 15-amp circuit breaker or fuse, and so on up the line.

The permissible loads according to circuit size are referred to in Code book Article 210-23(a), (b) and (c).

The load is never allowed to exceed the circuit rating. The three kinds of circuits are outlined in Sub-sections (a), (b) and (c). The definition and rules are a little tricky to understand. Let's look at them one at a time.

As we examine the Permissible Loads, keep in mind that these circuits have **two or more loads**. They are **not individual branch circuits**.

For example, a 20-ampere branch circuit can have lighting, appliances, or a combination of both attached. You could even attach a fixed appliance.

Individual branch circuits are covered later in this chapter and in Code book Article 422-5.

Remember the definitions of portable, stationary and fixed appliances? The fixed appliances are built in, the stationary are large but removable, and the portable are small plug-in types. Each type has Code rules.

15- AND 20-AMP BRANCH CIRCUITS

A 15- or 20-amp branch circuit can supply lights and appliances. However, the rating of any one portable or stationary appliance cannot exceed 80 percent of the amp rating.

In a 15-amp circuit, that converts to 1800 watts. In a 20-amp circuit, that converts to 2400 watts. This means you cannot put more than 1440 watts on a 15-amp circuit or 1920 watts on a 20-amp circuit. Actually, the homeowner does as he pleases, but that is the maximum load allowed by the authorities.

When you wire a fixed appliance to a 15- or 20-amp circuit, you are not allowed to exceed 50 percent of the circuit rating.

This means a 7.5-amp appliance on a 15-amp circuit and a 10-amp appliance on a 20-amp circuit. For example, a room air conditioner should operate at no more than 900 watts on a 15-amp circuit or at no more than 1200 watts on a 20-amp circuit.

At any rate, you are to use #14 or #12 wire; and if you wire fixed loads or prepare for a stationary load, be sure not to exceed the circuit rating.

Most of the wiring is easy and straightforward, causing few or no complications. The circuits in our example home which can be handled with 2400 watts each are the light, receptacle, small appliance and laundry.

Each light-receptacle circuit can handle 500 watts. The two small appliance circuits are 1500 watts and the laundry circuit is 1500 watts. These circuits lie well within the 80 percent ruling for portable and stationary appliances.

THE 25- AND 30-AMP BRANCHES

A 30-amp branch circuit at 120 volts can handle 3600 watts. This handling capacity is seldom if ever taxed. If it is expected to be, then #10 wire and a 30-amp overcurrent device should be used.

More than likely a 30-amp branch circuit is a 240-volt type. It can handle 7200 watts. The Code says that, if you attach portable or stationary appliances to this circuit, you cannot exceed 80 percent of the 7200 watts, which is 5760 watts. This 5.8-kilowatt load is of an appreciable size and is a common one in homes.

The Code does not apply the 50-percent limitation on fixed appliances to these circuits. These circuits are permitted to draw the maximum rating of the circuit. To a 30-amp branch circuit rated to carry 7200 watts, you can attach a fixed appliance with a nameplate rating of 7200 watts—the full 30 amps which the #10 wire is rated to handle.

Look up #10 wire in Table 310-16 of the Code. You'll see, among such common conductors as THW, that #10 wire is rated at 30 amps.

Notice that the table has no reference to 120, 240 or any voltage. The #10 wire is allowed to handle 30 amps, no matter what the voltage is. If the voltage is lower, the wire can't handle as many watts. If the voltage is higher, the wire can handle more watts, but the current remains the same.

It is worthwhile to study that last paragraph, since there is considerable confusion about the relationship between voltage and the current rating of a wire. Remember! the voltage doesn't change the current rating of a wire.

40- AND 50-AMP BRANCH CIRCUITS

If you look at Article 210-23(c) of the Code book, you will not find any percentage limitation on the circuit. Whatever wattage the circuit can handle is the limitation. The wattage does not have to be reduced by an 80- or 50-percent figure.

A 40-amp circuit uses a #8 conductor, and a 50-amp circuit uses a #6. Invariably, a residential circuit which carries such large amounts of current is a 240-volt circuit.

A 40-amp circuit at 240 volts can handle 9600 watts. A 50-amp circuit can handle 12,000 watts. If you have an air conditioning system with a nameplate rating of 12 kilowatts, you can safely use a 50-amp, #6 wire circuit. If the load is 9.6 kilowatts, use a #8 wire. It is approved by Code.

When you attach a motor or other large electrical load, you might have to change the wire size to a larger gage, but the ruling will come from the special load condition. For routine dwelling installations, no such consideration is needed.

These wire sizes of 80 percent, 50 percent, etc., in the different amp circuits is a confusing area. This is because there are three kinds of appliances, two kinds of circuits and two kinds of voltages.

ACTUAL CIRCUIT SIZING

Article 210-23 (Permissible Loads) of the Code book starts off with: "In no case shall the load exceed the branch circuit ampere rating." This seems like an obvious statement.

Actually, the dwelling wiring job is designed to accommodate to the expected load. You know what the load is going to be. According to the requirements of the load, you install a system. The load comes first, not the circuit. The circuit is sized to suit the load.

In our example home, we are given the total load. The feeder conductors are sized to accord with that total load.

The total load is the sum of all the individual loads. Each branch circuit is sized to supply its own load safely. In our dwelling there are 11 types of loads. This means that there will be 11 types of branch circuits.

Eight of the branch circuits are individual circuits: dishwasher, trash compactor, garbage disposer, water heater, range, dryer, air conditioner and heater.

Three of the branch circuits are composed of more than one circuit. The small appliance circuit has two component circuits, the lighting-receptacle circuit may have as many as six component circuits, and the laundry circuit several. These three branch circuits might contain a total of as many as nine component circuits.

LIGHTING-RECEPTACLE SIZING

This dwelling could need as many as 60 outlets all told. Since 10 outlets on one branch is average, this house will require six branch circuits (Fig. 13-9).

If the branch circuits are all 15-amp circuits, or if they are rather evenly divided between 15- and 20-amp circuits, the panel box will need fuses or circuit breakers of 15- and 20-amp ratings.

Each 15-amp circuit will need #14 wire and each 20-amp circuit will need #12 wire. Try to find a good price on #12 wire, since this gage of wire is ideal for circuits of 15 and 20 amperage. The #12 wire is used for the phase line as well as for the neutral line.

The circuits are all 120-volt lines. They are installed as two-wire circuits with equal current flowing in the phase and neutral lines.

It is interesting that six 15-amp circuits can draw as much as 90 amperes. This converts to 10.8 kilowatts. This is considerably more than the 5.1 kilowatts that we calculated the feeder would need to supply the six circuits. However, the possibility of all of the circuits being operated to capacity at the same time does not exist.

SMALL APPLIANCE AND LAUNDRY

The six lighting-receptacle circuits, the two small appliance circuits, and the laundry circuit are all available to the homeowner. The homeowner is able to plug an appliance of any size into these circuits whether the 80-percent rating is exceeded or not. No one can police this homeowner option. It is the wireman's job to protect the circuit with the correct fuse or circuit breaker.

The small appliance and the laundry circuits are all of 20-amp rating and operate in the 120-volt lines. Thus, #12 wire is used in these circuits (see Fig. 13-9).

To protect the nine circuits of our example home, nine single-pole overcurrent devices will be installed in the panel.

INDIVIDUAL CIRCUITS

The lighting-receptacle, small appliance, and laundry circuits are not individual circuits. They each contain two or more receptacles. Even the laundry circuit can have a couple of duplex receptacles on a 20-amp line.

These circuits are more or less covered in Code book Article 210-23 (Permissible Loads). Individual appliance circuit are covered in Article 422—Appliances under B. Branch Circuit Requirements. Notice that 422-5 (b) is designated Circuits Supplying Two or More Loads. It refers you to 210-23. That is where you get into all of that 80- and 50-percent calculation of portable, stationary, and fixed loads.

In Article 422-5 (a) (Individual Circuits) you are told that the rating of a branch circuit shall not be less than the nameplate rating of an appliance or other load. This means you can attach an appliance with a rating equal to the limit of a wire's ampacity. If you have a 50-amp circuit of #6 wire, you can attach a range drawing 50 amps.

In our example home, individual circuits are required for the dishwasher, trash compactor, garbage disposer, range, water heater, dryer, air conditioner and electric heater.

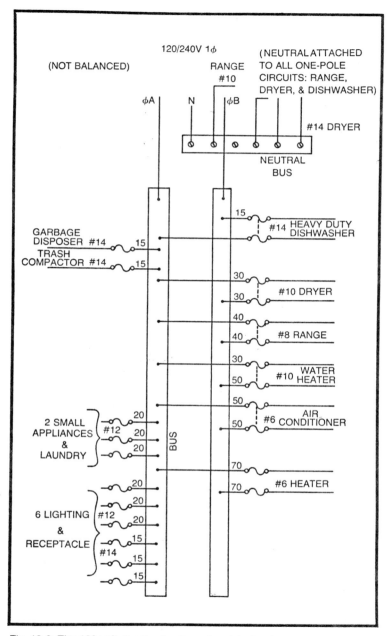

Fig. 13-9. The 120-volt, 1-pole circuits and certain 2-pole appliances attach to neutral. (This panel layout is for illustration purposes only. It is not balanced properly.)

DISHWASHER, TRASH COMPACTOR AND GARBAGE DISPOSER

The dishwasher has a nameplate rating of 2.5 kilowatts. It is an unusual 240-volt type since it is large. An appliance which consumes more than 1625 watts is better served with 240 volts than 120 volts. Under 1625 watts, a 120-volt supply is the more efficient.

Even though many appliances are in excess of 1625 watts, they are operated as plug-in types in #12 wire, 20-amp circuits. It is more convenient. You'll find barbecues around 1800 watts, and room air conditioners at 1900 watts. However, the plug-in capability outweighs the electrical efficiency factor. These appliances would operate more efficiently at 240 volts even though they are rated at 120 volts.

Our dishwasher (Fig. 13-9) at 2500 watts (2.5 kilowatts) has a 240-volt input. $I = P/E = 2500/240 = 10.4$ amps. A #14 wire of 15-amp capacity is permissible. It would be better, however, to install a #12 wire in anticipation of additional loads in the future.

The trash compactor uses 1000 watts. It is an individual circuit and load draws $I = P/E$. The voltage here is 120 volts. It is less than the 1625-watt crossover point to 240 volts.

$$I = \frac{1 \text{ kW}}{.240} = 4.1 \text{ amps}$$

The garbage disposer presents an identical load to its branch circuit. At 1 kilowatt it also draws 4.1 amps. Both circuits are easily served with a #14 wire and 15-amp fusing.

The dishwasher needs a two-pole overcurrent device. It has two phase lines since it uses 240 volts. The poles are 15 amps each (or 20 amps if #12 wire is used).

The trash compactor and garbage disposer use single-pole overcurrent. The three appliances occupy four spaces in the panel: two for the dishwasher and one each for the other two appliances.

DRYER

Both the range and dryer use individual circuits of 240 volts. Both draw considerable current. Both have a neutral wire in addition to the two phase wires.

Overcurrent protection is needed in the phase wires. Each of these appliances occupies two spaces in the panel. The two neutrals are attached to the neutral bus bar in the panel (Fig. 13-9).

The dryer has a nameplate rating of 6 kilowatts. It is a 120/240-volt appliance. The heating elements and tumble motor use 240 volts while the light and timer use 120 volts.

Whatever current is used in the 120-volt circuit is returned through the neutral wire. No neutral current flows from the heating element.

$$I = P/E = \frac{6 \text{ kW}}{.240} = 25 \text{ amps}$$

A 30-amp individual circuit is strung and two #10 wires are used for the ungrounded sides. #14 wire can be used as the neutral return. The number of amps in the neutral is less than five.

RANGE

The range circuit is like the dryer in that it is also 120/240 volts. The range's light, timers and other small gadgets must be operated on 120 volts. The heating elements use 240 volts.

The dryer, whenever it is operating, is drawing current to its full capacity, but the range rarely so. As a result the Code lets you demand the range circuit wire, just as you demanded the range feeder capacity. Table 220-19 can be used. The permission is granted in Article 422-5 (a), Exception No. 3.

Thus, you can consider the load as 9.2 kilowatts instead of as the nameplate rating of 15 kilowatts.

$$I = \frac{9.2 \text{ kW}}{.240} = 38.3 \text{ amps}$$

You can use a 40-amp circuit, which means two #8 wires. A smaller neutral can be used. The rule of thumb is to use one size smaller for the neutral, in this case a #10.

The range can have its own separate input from the panel or be one of the panel group.

WATER HEATER AND AIR CONDITIONER

Neither the water heater nor the air conditioner uses a neutral wire. The circuits are straightforward for the 15-kilowatt water heater and the 12-kilowatt air conditioner. They both are two-pole appliances and each occupies two spaces in the panel.

$$\text{Water Heater } I = \frac{5 \text{ kW}}{.240} = 20.8 \text{ amps}$$

$$\text{Air Conditioner } I = \frac{12 \text{ kW}}{.240} = 50 \text{ amps}$$

The water heater will need a 30-amp circuit since it draws more than 20 amps. Two #10 conductors connect it to two places on the panel.

The air conditioner needs a 50-amp circut with two #6 conductors.

HEATER

The electric heating operates at 15 kilowatts. The conductors to the panel are 240 volts.

$$I = \frac{15 \text{ kW}}{.240} = 62.5 \text{ amps}$$

This requires a heavy duty copper wire to carry the current drain. Code book Table 310-16 shows that type THW requires a #6 conductor.

There is no neutral, so only two #6 conductors are needed.

THE PANEL LAYOUT

The panel has to be rated according to the feeder ampacity, not according to the branch circuits. It must have enough spaces to hold all the fuses or circuit breakers, but it does not have to be able to carry the branch currents.

If all the branches of our example house were operating at the same moment, and each load was on full, the wattage would be 67.1 kilowatts. The current would be 280 amps.

The panel is sized at 150 amps. It actually is required to handle only 128 amps after the demanding. That is a long way from 280 amps.

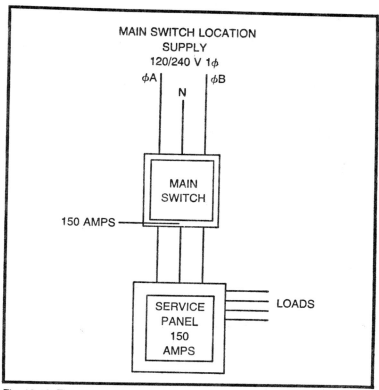

Fig. 13-10. The main switch is installed on the supply side of the circuit so the panel can be disconnected.

The rating of the panelboard is determined by the size of the copper bus bars to which the fuses or circuit breakers are attached. The bus bars must not be smaller than the computed load as described in Code book Article 220.

The panel might or might not house the main circuit breaker. When the main breaker is not in the panel, it will be positioned elsewhere in the system. The main switch in a home should be situated on the supply side of the fuses rather than on the load side. This is to allow the entire panel to be disconnected from the supply so that fuses can be replaced with safety. If the switch were situated on the load side, the panel could not be disconnected during fuse changing (Fig. 13-10).

A panel is not permitted to contain more than 43 fuses or circuit breakers. A single-pole breaker is counted as one device, a two-pole breaker as two devices, and a three-pole breaker as three devices.

The breakers used are classified as CTL types. The CTL stands for "circuit limiting." The breakers are plug-in types and the breaker receptacle is called a stab receiver.

The stab receivers are shaped either as F slots or as E slots. The F slots accept one-pole breakers and the E slots accept two-pole breakers. There are never more than 42 available slots. The main breaker is not counted among these.

In our example house, we need (in addition to the main breaker) the following breakers:

Lighting & Receptacle	3-15A
Lighting & Receptacle	3-20A
Small Appliances (2)	2-20A
Laundry	1-20A
Dishwasher	2-15A
Trash Compactor	1-15A
Garbage Disposer	1-15A
Dryer	2-30A
Range	2-40A
Water Heater	2-30A
Air Conditioner	2-50A
Heater	2-70A
	(See Fig. 13-9)

Rural Wiring

In the life of the average American farm, electricity plays a vital role. Not only does it figure importantly in making life comfortable in the farmhouse, electricity milks cows, increases egg production, cures harvested crops, and runs machinery. The list goes on and on. Electric power is the difference between a modern profitable farm and a losing proposition.

FARM NEEDS

The needs are special when compared to electrical needs beyond the farm. The Code book devotes Article 220-40 to farm loads (Fig. 14-1).

A farm typically has a dwelling, barns, chicken coop, curing and storage sheds, machinery sheds, stock shelters, and occasionally line shacks.

The electric power is the same as in the city since it comes from the utility company, but there are some major differences in distribution and grounding. Distances between load sites are usually greater and there is seldom available a water pipe system as in the city for attaching a safe ground to.

A farm is a business and when the installation is planned, growth is one of the considerations. If rapid growth is hoped for, then the capability of expanding the electrical needs should be considered.

Method for Computing Farm Loads for Other Than Dwelling Unit

Ampere Load at 230 Volts	Demand Factor Percent
Loads expected to operate without diversity, but not less than 125 percent full-load current of the largest motor and not less than the first 60 amperes of load............100	
Next 60 amperes of all other loads 50	
Remainder of other load...25	

220-41. Farm Loads—Total. The total load of the farm for service-entrance conductors and service equipment shall be computed in accordance with the farm dwelling unit load and demand factors specified in Table 220-41. Where there is equipment in two or more farm equipment buildings or for loads having the save function, such loads shall be computed in accordance with Table 220-40 and may be combined as a single load in Table 220-41 for computing the total load.

Method of Computing Total Farm Load

Individual Loads Computed in Accordance with Table 220-40	Demand Factor Percent
Largest load 100	
Second largest load...75	
Third largest load..65	
Remaining loads	

To this total load, add the load of the farm dwelling unit computed in accordance with Part B or C of this article.

Fig. 14-1. These two tables from the National Electrical Code book are devoted to the calculation of farm loads.

Extra capacity or provision for the easy adding of extra capacity is important.

There are many types of motors on farms which are never found in urban residences. For example, a dairy barn can have several 7½-hp motors, sheds a 5-hp motor and wells a 10-hp motor. The wiring can run for hundreds of feet.

The electrical needs of farms are different and important.

THE YARDPOLE

Near the farm dwelling is a service drop from the utility company. It enters a weatherhead that feeds into the main panel. Between the weatherhead and the panel is the kilowatt hour meter. The panel distributes the branch circuits throughout the dwelling.

In an apartment house the situation is similar, except that all of the apartments are fed from one service drop. Each apartment has its own meter and panel.

A farmstead consists of many individual buildings just as an apartment house consists of many separate apartments, but the farmstead needs only one meter. Yet the farm needs a number of panels, one for each building, just like the multi-unit apartment house. What is the best way to handle this different kind of feeder arrangement? To complicate the situation, the farm buildings are not clustered. In fact, they're often spread widely.

The yardpole (Fig. 14-2) solves the problem. With the yardpole centrally located, the following circuitry can be installed.

A three-wire, 120/240-volt, single-phase service line can be run by the utility company to the top of the yardpole. The three wires drop to three insulators on the pole, one above the other. The top insulator receives neutral and the bottom two insulators receive the A and B phase lines.

A weatherhead is installed above the insulators. The weatherhead has 5 holes in it, even though there are only three wires involved. You'll see the reason for the two extra holes in the next section.

The three wires are run into the weatherhead. A conduit is installed from the weatherhead to a meter mounted on the pole at eye level. The three wires are run to the meter.

Below the meter can be a weatherproof main switch which will cut off the entire farm, if need be. The two phase lines are run to the main switch. The Code does not require the switch, but its use is advisable.

YARDPOLE FEEDERS

The neutral runs to the switch box. It attaches to the neutral bus bar in the switch box. The neutral ends there.

The two phase lines from the switch are then run back up the pole through the meter and conduit. These feeders from the three insulators run down the pole, through the meter and switch, and then back up the pole. They are the main feeders. All of the farm current has to pass through these feeders. They must be sized accordingly. The neutral only makes the trip down the pole. The two phase lines make the trip down and then back up through the same

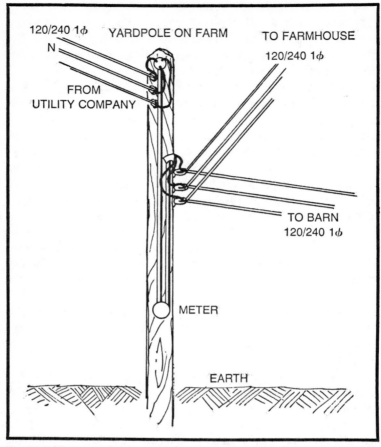

Fig. 14-2. The meter for the entire farm normally is installed on the yardpole below the service drop.

conduit. That is the purpose of the two extra holes in the weatherhead: for the entrance of the two returning phase lines. Therefore, the weatherhead has three lines, two phase and one neutral, going in, and two lines, both phase, coming out.

The two phase lines coming out are then run to other sets of insulators. The other sets of insulators, one set for each farm building, are also attached in a vertical arrangement, with neutral on top.

Three-wire subfeeders are then run from each set of insulators to a farm building. Since each building has its own main switch, the Code doesn't require a switch on the yardpole.

These subfeeders to each building carry current nowhere but to it. The subfeeders are sized accordingly and are therefore smaller than the main feeder. That is, as far as electricity goes.

If the wire run gets too long, larger size subfeeder wires might be needed for their greater tensile strength and to reduce voltage drop. Most of the time, the current-carrying capacity of the wire is satisfactory.

SINGLE OR DOUBLE STACKS

The term "stacks" is used in connection with the weatherhead. A weatherhead with five holes is a single stack. Two weatherheads with three holes are a double stack.

The weatherhead described in the previous section was a single stack (Fig. 14-3). A double stack (Fig. 14-4) uses two conduit lengths with the two weatherheads. One conduit carries the three wires down the pole. The second conduit carries the three wires back up the pole. The neutral is not ended at the meter site.

The neutral of the single stack is jumped from the power line neutral directly to the subfeeders. The neutral of the double stack runs alongside the phase lines.

One advantage of the double stack is that cable (as opposed to conduit) can be used for its installation. In the single stack arrangement, there are five wires in the pipe, two phase lines and a neutral running down and two phase lines running up.

With a double stack, only one cable runs down and only one cable returns. This can be an inexpensive way to wire the pole.

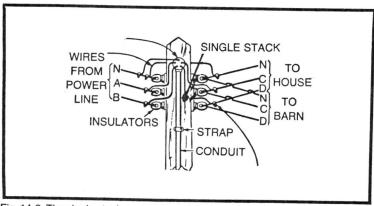

Fig. 14-3. The single stack uses a single length of conduit for containing both the rising and descending wires.

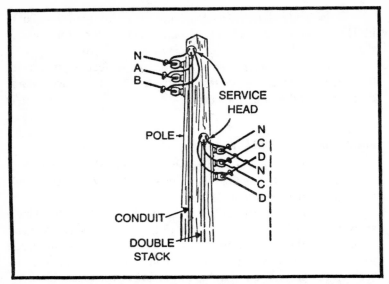

Fig. 14-4. The double stack uses two lengths of conduit and two weatherheads. One conduit houses the rising wires, the other conduit the descending wires.

GROUNDING THE YARDPOLE

Since there are no waterpipes networking throughout the farm, as they do throughout the city, grounding must be done mostly by ground rod. The ground rod is not as good a ground as the urban water pipe system, but if installed correctly is satisfactory.

A farm has chemicals, manure, poor drainage and other materials which abrade and deteriorate electrical system hardware. These normal farm materials can eat away conductors and rods unless care is taken.

The best ground source is the neutral on the pole coming from the power company. This neutral is already being run to the meter and switch to put them at zero volts. It is up to you to run another grounding conductor separate from the neutral in the meter to earth.

A direct grounding conductor is run from the utility company neutral atop the pole to earth. It must be run outside the conduit. Some local codes allow the ground wire to be run alongside and in contact with the conduit.

Either way, the actual ground rod should be located about two feet away from the pole. A trench is dug and the ground rod is driven into the bottom (Fig. 14-5). The ground clamp is the only part of the rod left exposed in the bottom of the hole.

The trench is a bit more than two feet long, extending between the yardpole and the rod. The grounding conductor runs directly to the bottom of the trench and makes a right-angle turn to the rod. After a good solid solderless connection is made at the ground clamp, the trench is filled and packed tightly to protect the wire, rod, and clamp from the farm chemicals.

GROUNDING THE BUILDINGS

Each building whose switch receives a subfeeder gets a ground rod. The ground rod and grounding conductors should be trenched in the same manner as the pole.

All of the ground rods should be tied together. If there is a farm water system with piping, it should be attached to the ground system, too.

The more surface area you have touching earth, the safer the grounding system will be. You cannot have too many grounds.

The Code Article 250-83 gives the rules about ground rods. They shall not be less than eight feet in length. They must have a resistance between rod and ground of no less than 25 ohms.

This means, if there is a voltage of 120 and the resistance is 25 ohms, that five amps will be passing through the grounding wire (Fig. 14-6).

BUILDING FEEDERS

Each service entrance to each farm building is not much different from the service entrance to other buildings. Even if it doesn't have a meter, each building has a switch and panel.

Each subfeeder is the feeder for that building. It is sized to carry the current which is used in that building. If the building is the

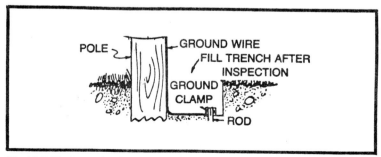

Fig. 14-5. The ground rod is installed in a trench about two feet from the yardpole. The trench is later filled with dirt.

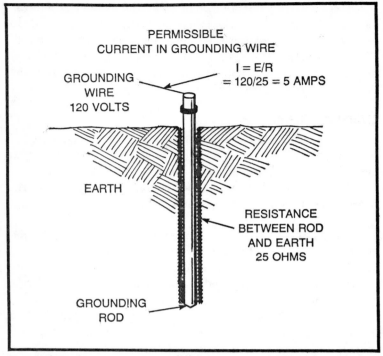

PERMISSIBLE
CURRENT IN GROUNDING WIRE

GROUNDING
WIRE
120 VOLTS

$I = E/R$
$= 120/25 = 5$ AMPS

EARTH

RESISTANCE
BETWEEN ROD
AND EARTH
25 OHMS

GROUNDING
ROD

Fig. 14-6. The Code permits the resistance between earth and the grounding rod to be as high as 25 ohms.

farmhouse, the load is computed exactly as it would be for any dwelling, according to Article 220, Part B or Part C.

If the building is a dairy barn, the building's load is computed by figuring the lighting, convenience outlets, milking machine, feed grinder, motors, fans, hay driers and so on. The demanding is calculated for the feeder according to Article 220-40 (Farm Loads).

All of the other buildings are computed in a similar way, according to their particular needs. Chicken coops have critical lighting needs to keep the hens contented. Storage sheds need special environments. Machine sheds need special circuits for the operation of special equipment.

All of the buildings, once their individual loads have been computed, will have their own feeder wires and service of the appropriate size.

After all of the buildings have been computed, the total farm load can be added together and demanded.

Table 220-40 assumes an amp load powered by 230 volts. (This is considered as 240 volts, too.) Table 220-40 gives you the demand factors to be used when you calculate each building.

Once you know the demanded load of each building, you can refer to Table 220-41. This table gives you the demand factors for the total farm load. The total farm load determines the size of the feeder which runs up and down the pole.

USE NONMETALLIC MATERIALS

The electrical system on the farm gets sprayed on, excreted on, and has a ground which is less effective than a city water pipe ground. Steel conduit and fittings are easily corroded and can crumble. The ground rod only has ten feet of length for making contact with earth. The city water pipes have miles and miles of length. The more pipe in contact with the earth, the better and safer the ground system.

If the pipe or a fitting should break, some of the neutral's current could escape. A segregated piece of pipe which is not continuous with the rest of the grounding system could become live. A human being could be shocked or an animal killed. Cattle have a much lower tolerance to electric shock than humans do and perish easily from shocks that would do no more than jolt a human.

For these reasons, nonmetallic sheathed cable, nonmetallic boxes, and other plastic materials should be used in place of a metal conduit system.

"Barn cable," an NMC type, is the best variety of conductor to use. It can be used throughout the installation and is even recommended for use in the farmhouse.

Nonmetallic boxes are available with covers and mounting ears and can be nailed directly to wooden structures. The plastic boxes are not tightened down like metal boxes, because plastic is brittle and will not flex to the degree that metal will. A little timber movement and a tightly mounted plastic box would crack. A small amount of slack is given during the installation to avoid this complication.

Porcelain or Bakelite sockets are the best kind to use on the farm. Humans and animals can safely rub against any of these insulated cables and boxes and come away unscathed.

FARM LOAD CONSIDERATIONS

When the electrician undertakes a farm wiring job, he'll find himself engaged in some long conferences with the farmer. The electrician will need the farmer's advice on where to place the lights, outlets, and switches in the outbuildings. The farmhouse is just like any other dwelling except for the needs of the farmer as an individual. He might perform in the dwelling some type of chore which requires a special circuit.

The electrician should be prepared to offer suggestions, such as about switches that can be operated by an elbow when the arms are full, or about lighting that would cast no shadows in an area occupied by hazardous equipment. When the farmer speaks about sun lamps in the poultry house or the nocturnal duties which require outdoor illumination, the electrician should adapt the installation accordingly.

Typical farm buildings use about as much electricity as homes. There are small buildings such as small poultry houses which might need only one circuit. There are large buildings which might require a couple of dozen circuits.

In the small building with one circuit serving two overhead lights and a receptacle, all you need is one toggle switch. This meets Code specifications since the building's main switch is capable of disconnecting both the receptacle *and* the lights. If a constantly live receptacle is needed, then the electrician will have to install a second circuit and a toggle switch.

He can use toggle switches or circuit breakers as the building's main switch, but the Code specifies that no more than six of these devices can be installed.

In a building with two dozen circuits, the electrician will need to install a large main switch for disconnecting all the circuits. He could use any number of switches up to six but one is the most practical in such a building.

FARM BUILDING LOAD CALCULATION

The Code dictates how to demand a farm building. This calculation allows the electrician to determine the size of the subfeeders which run from the yardpole to various buildings (Fig. 14-7).

The yardpole is located centrally and usually close to the largest load. This location offers the least length of cable for the largest load. The shorter the length of heavy cable, the less the expense. If a long run of cable is needed, it should be connected to the lightest load, if possible.

The yardpole's proximity to the largest load keeps the voltage drop low, too. This saves energy. The dwelling load is not calculated

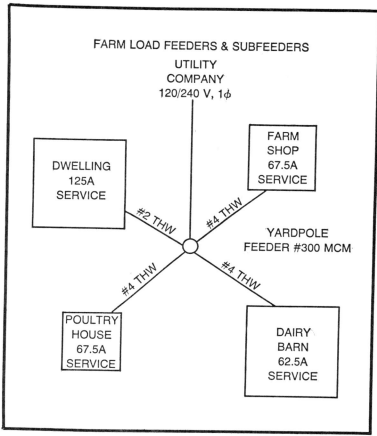

Fig. 14-7. Each farm building has its own particular needs, which are dependent on the functions it performs.

by the procedure in Section 40 of Code book Article 220. Instead, it is calculated by Parts B and C of the same article.

Table 220-40 computes the loads of individual buildings. The first line of the table is hard to understand. It has to do with the first 60 amperes of the load. There is no demanding of the first 60 amps. It is calculated at 100 percent.

Because the language in that sentence is hard to understand, the meaning is best explained with an example.

Take the calculation of the circuits in a farm shop. The shop will use 10 amps for lighting, 15 amps for power tools, and 40 amps for a motor.

Lights	10 amps
Tools	15 amps
Motor	40 amps
Total	65 amps

Lookin 〈 close at the confusing sentence, the lights and tools are added okay, but the motor has to be multiplied by 125%. That changes the 40 amps to 50 amps (40 × 1.25 = 50). Then the three kinds of loads add up to 75 amps.

The sentence orders the first 60 amps at 100 percent, but permits the next 60 amps to be demanded at 50 percent. But there aren't 60 amps more; there are only 15. That adds 7.5 amps to the original 60 amps. The total load for the farm shop is 67.5 amps.

This means that the feeder conductors have to be able to carry 67.5 amps. We refer to Table 310-16 and find that a THW #4 copper will do the job for the farm shop. (Fig. 14-7).

POULTRY HOUSE LOAD

To compute the ampere load of the poultry house, we again use the procedure prescribed in NEC Table 220-40. We determine that 40 amps will furnish power enough to energize the ordinary outlets, the brooder outlets, ultra-violet lamps, germicidal lamps, and all incandescent and fluorescent lighting fixtures.

Appliances such as water warmers, egg cleaners, candlers, graders, scalders, waxers and other egg production equipment will add another 10 amps to the load.

The poultry house usually has at least one motor around and it could draw 20 amps.

The poultry house total load adds up like this.

GIVEN LOAD		COMPUTED LOAD
Lighting	40 amps	40A
Appliances	10 amps	10A
Motor	20 amps × 1.25 =	25A
		75A

It turns out that the poultry house needs the same size feeders as the farm shop, even though the circuit distributions of the two are entirely different.

320

Sixty amps are demanded at 100 percent and 15 amps at 50 percent for a total of 67.5 amps.

As feeders into the poultry house, #4 THW conductors can be used.

DAIRY BARN LOAD

The dairy barn has a load of 20 amps for lighting and outlets, another load of 20 amps for small appliances, and a third load of 20 amps for a motor.

GIVEN LOAD		COMPUTED LOAD
Lights	20 amps	20A
Appliances	20 amps	20A
Motor	20 amps × 1.25	25A
		65A

The first 60A at 100%	=	60.0A
The next 60 A at 50% (5A)	=	2.5A
		62.5A

Here again, there isn't much difference between this dairy barn load and the loads of the other two buildings. The same size service is used in the dairy barn, too.

CALCULATING THE TOTAL FARM LOAD

After all of the individual services have been computed, and all of the subfeeders from the yardpole sized, it is time to calculate the size of the feeder that runs up and down the pole between the utility company's drop and the subfeeders. The total farm load passes through this pole feeder.

As mentioned earlier, a main switch, fuse or circuit breaker is not required for this service feeder. Its use is recommended, though, for cutting off the entire farm from one place, if the need should ever arise. Otherwise, it would be necessary to go to each building to disconnect those loads. If there is no main disconnect, the feeders will always be live and the subfeeders can't be worked on when needed.

However, working on the subfeeders is a lineman's type of job and is best left for the lineman. A lineman is well prepared to work on live wires. Furthermore, the switch on the pole is quite expensive, especially on large farms, and the power consumption rivals that of an industrial plant.

Calculation of the total load for this average size farm is done in accordance with the models in NEC Tables 220-40 and 220-41. Table 220-40 deals with the individual building loads and Table 220-41 computes the total load.

Table 220-41 does not have anything to do with the farm dwelling. The dwelling is computed like any non-farm dwelling and is then added to the total load. Table 220-41, as it lists largest load, second largest load, etc., is referring to the outbuildings only, not to the farm dwelling.

A note in small print beneath the table makes the dwelling calculation clear.

Our example farm has a total of four buildings, three work buildings and the dwelling. We use Table 220-41 as a model and arrange our farm's loads with the largest first, the next largest second, and so on.

Farm Shop	67.5A Largest Load at 100% =	67.5A
Poultry House	67.5A Second Largest at 75% =	50.6A
Dairy Barn	62.5A Third Largest at 65% =	40.6A

Total Farm Load (excluding dwelling)	= 158.7A
Dwelling	125.0A
Total Farm Load	283.7A

Table 310-16 prescribes #300 MCM copper for the power pole feeder when the pole has no switch or overcurrent device.

FARM FEEDER MINIMUMS

On small farms the feeder can be smaller than #300 MCM. Under no circumstances, though, should the feeder on the yardpole be smaller than #2 wire.

322

For the subfeeders which extend between the yardpole and individual buildings, conductors no smaller than #10 gage must be used when the span is less than 50 feet. When the span is greater than 50 feet, then #8 gage is the smallest size permitted.

The reason for the above prescriptions is mechanical, not electrical. The overhead wiring is subject to snow, hail and wind. It must have a tensile strength sufficient to withstand the elements.

Theory of Services

A wireman cannot consider himself a complete electrician until he fully understands how and why things operate the way they do. This *how* and *why* represent the theory of electricity. This chapter focuses on the theory of service. The concepts aren't always simple, but once you grasp them, you'll be a better—and more confident—electrician, whether professional or do-it-yourselfer.

TRANSFORMERS

A transformer is able to change current from one voltage to another. If it is a low voltage transformer, it is able to change 120 volts to 10 volts. It is able to change 480 to 240, 4000 to 240 or perform whatever change you'd like.

The transformer can't make this change with DC, only with AC. Fortunately, most electric power is in the AC form and can be changed.

Otherwise, it would not be feasible to transmit electric power over long distances. If you had to send thousands of kilowatts at 240 volts over miles and miles of cable from the generating plant to a consumer, the cable would have to be so thick, in order to avoid voltage drops, that the cost would be out of the question.

However, very high voltages can be transmitted through conductors of modest and economically feasible diameters. With a transformer located near the consumer, the high voltages can be reduced to the convenient 240 volts on which our electrical system is based (Fig. 15-1).

A transformer has no moving parts. It consists of three sections: (1) primary windings, (2) secondary windings, and (3) the transformer core of magnetic material. None of these sections move. The electric current and the resulting magnetic field do the moving.

Figure 15-2 illustrates the basic transformer. The high tension line supplies the primary winding of the transformer with a 2400-volt input. As the alternating current courses back and forth in the primary, a magnetic flux is produced around the windings. The soft iron core retains the magnetic flux.

The flux expands and contracts as the AC passes from zero to maximum voltage and back. During expansion the flux from the primary cuts through the secondary windings and generates a voltage in the secondary.

According to the number of windings in the primary and secondary, the voltage generated in the secondary varies.

TRANSFORMER RATIOS

The 1:1 transformer has the same number of coil windings in the primary as in the secondary. Therefore, the same voltage that is in the primary is generated in the secondary. What use does the 1:1 transformer have? The 1:1 transformer physically separates the primary circuit from the secondary circuit. The windings are also electrically separated, but they are magnetically coupled. This arrangement is useful when DC levels in the primary have to be kept away from the DC level in the secondary. This is an isolation type of transformer (Fig. 15-3) and it has its special uses.

An electrician will be called upon to wire transformers with different ratios. For example, a typical three-wire, 120/240-volt service gets its voltage from the secondary of a 10:1 transformer.

The primary of a transformer consists of the windings to which the supply voltage is applied. The secondary of a transformer consists of the windings that apply the voltage to the line load.

It is known that when the primary coil of a transformer has ten times as many turns as the secondary coil, then the primary will have ten times as much voltage as the secondary.

The pot on the utility pole could be a 10:1 transformer. It could have ten times more windings in the primary than in the secondary. The primary is supplied by the generating plant with 2400 volts. The

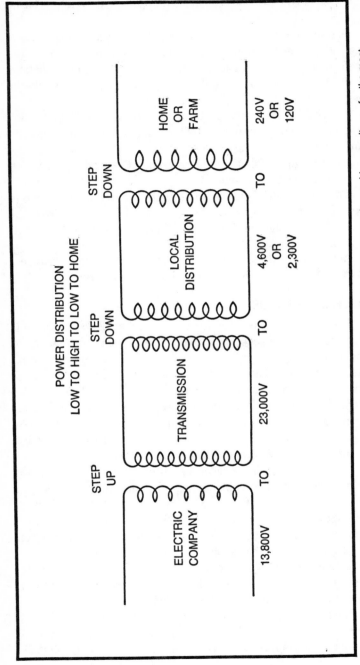

Fig. 15-1. A typical electric company power distribution system uses transformers to raise and lower voltages for the most efficient transmission of current.

secondary will then be able to produce 240 volts. This is called a step down transformer.

If you reverse the path of the current in the 10:1 transformer by attaching the secondary windings to the incoming supply, then you have converted the transformer into a 1:10 step up transformer.

When the secondary is attached to the 2400 volts, it becomes the primary. The smaller winding then gets the 2400 volts. The larger winding develops a stepped up voltage of 24,000 volts.

Transformers can have any kind of ratio according to their use. Their importance lies in their ability to change voltage or current from one value to another. A 1:2 transformer doubles voltage, a 1:3 triples voltage, and a 5:1 produces one-fifth the voltage.

CURRENT AND WATTAGE IN XFORMERS

A good rule to follow when you're figuring the relationship between voltage and amperage in a xformer is that when the voltage

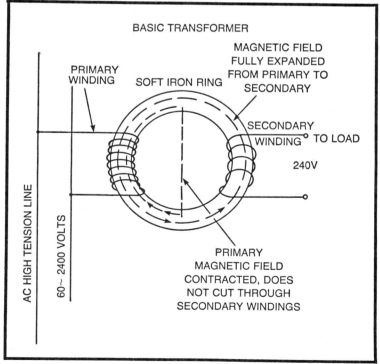

Fig. 15-2. The basic transformer has three parts: the primary, secondary, and core. The transformer is able to generate a secondary voltage on order.

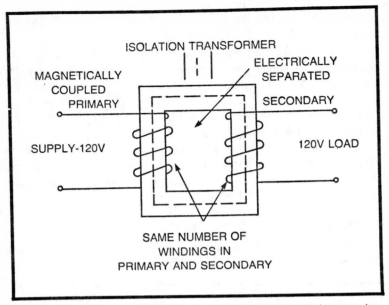

Fig. 15-3. A transformer that has the same number of windings in the secondary as in the primary has a turns ratio of 1:1.

change is a whole number, the amperage change will be in the amount of the fractional inverse of the whole number. For example, if the voltage is doubled (2), the amps are halved (½), and if the voltage is halved (½), the amps are doubled (2).

About wattage, don't concern yourself. The wattage in the primary is the same as in the secondary. If there is a 300-watt load in the secondary, then the primary is going to deliver 300 watts.

All this works out with the Power Formula and Ohm's Law, $P = EI$, $I = P/E$, and $E = P/I$.

For example, suppose you want to install three 100-watt pool lights. You use a step down 10:1 transformer that changes the 120 volts to 12 volts. The lamps are 12-volt types. What is the wattage in the transformer, in the secondary current, and in the primary current? (See Fig. 15-4.)

First of all, you assume that the xformer is 100 percent efficient.

The wattage is the total load. There are three 100-watt lamps, totaling 300 watts.

The secondary current is $I_{sec} = P/E_{sec}$
$I = 300/12V = 25$ amps in secondary

The primary current is $I_{pri} = P/E_{pri}$

$I = 300/120V = 2.5$ amps in primary

That is the ordinary way to figure out current and wattages in transformers.

VOLT AMPERES

At a fast glance, volt ampere would seem to mean volts times amps—which also equals watts. In other words, a VA is a watt. In lots of cases it is.

In DC circuits a VA is equal to the watt. In many AC circuits the VA equals the watt. These AC circuits are the ones that light lamps and produce heat (Fig. 15-5a).

The AC circuits that do not involve currents through windings to produce magnetic fields have their volt amps equal to their watts. They are called non-inductive devices.

On the other hand, when a current travels through a coil of wire (also called an inductance), the volt amps are greater than the watts (Fig. 15-5b).

The volt amp is the product of volts times amps. The watt is the actual work that is produced. If the watt is smaller than the volts and amps are capable of producing, there is a loss of work.

These losses occur in coils, transformers and other inductive type devices.

The original current and voltage is the volt amp. The resultant current and the voltage produce the actual wattage work. It is not a heat loss, it is a current loss.

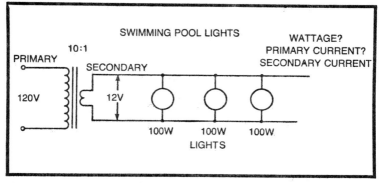

Fig. 15-4. A typical problem that an electrician encounters is the need to calculate the wattage and current values at which a transformer will be operating.

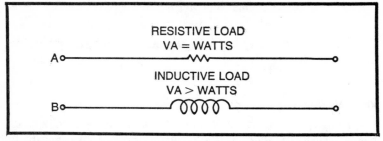

Fig. 15-5(a). In a resistance load, volt amps equals watts. (b). In an inductive load, volt amps exceeds watts.

The wattmeter doesn't record the volt amperage; it records the wattage. The consumer doesn't pay for the volt amperage; he pays for the actual wattage.

POWER FACTOR AND EFFICIENCY

The efficiency of the swimming pool xformer was assumed to be 100 percent. This meant there was a perfect transfer of voltage from primary to secondary without a loss of heat. Efficiency is one term.

Another term which refers to the relationship between watts and volt amps and *not to heat loss* is "power factor." The power factor, abbreviated PF, is a fractional figure, never larger than 1. The quantity 1 is called Unity Power Factor.

The power factor fraction is calculated from the formula

$$PF = \frac{watts}{volt\ amperes}$$

The power factor can be measured right in the circuit with a wattmeter and ammeter. Suppose you have a small motor circuit on a 120-volt line.

The wattmeter indicates that 240 watts are being used as 4 amps of current are being drawn. The power factor is

$$PF = \frac{watts}{volt\ amps}$$

$$= \frac{240W}{120V \times 4A} = .50$$

330

If the four amps of current were only two amps, then

$$\frac{240W}{120V \times 2A} = 1 \text{ (Unity Power Factor)}$$

This is assuming that the efficiency is 100 percent, no heat loss at all. In general, the larger the transformer, the more efficient it is. In gigantic transformers, the efficiency is greater than 99 percent.

KILOVOLT-AMP AND KILOWATT LOADS

Even though the kilowatt meter does not record kilovolt-amp loads, the utility company might be interested in measuring the kilovolt-amp loads of large installations.

If each of two factories uses 200-amp, 240-volt, single-phase service, they both need lines to supply 48,000 volt amps.

Suppose one has a power factor of unity or 1, and the other 0.5. One would use 48,000 kilowatts, but the other only 24,000 kilowatts.

The power company has to install the same wire drop, transformer, and generator capability. The power company only gets half as much money from the second factory.

Of course, the power company will not stand for this, so they will charge not only for the kilowattage but also for the kilovolt amperage.

A low power factor is usually corrected, since it wastes capability. Correction is done by means of special electrical equipment.

GETTING SERVICE FROM THE XFORMER

The service delivered by the power company comes from the secondary of a transformer. The transformers are wound to give two-wire, three-wire, four-wire, single-phase, and three-phase forms of current. The common types are,

1. 2-wire, single-phase, 120 volts
2. 3-wire, single-phase, 120/240 volts
3. 4-wire, 3-phase, wye, 120/208 volts
4. 4-wire, 3-phase, delta with centertap, 120/240 volts

All of these services can originate from a three-phase line system that comes from the utility company. Each kind of service requires a different transformer to supply the special kind of service.

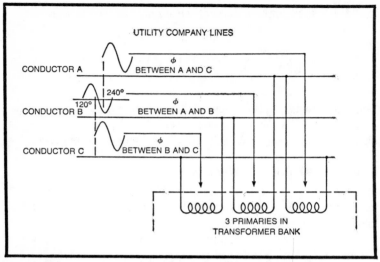

Fig. 15-6. Between any two wires of a 3-phase system is a single-phase current.

The utility company lines carry three phases. Each line does not carry a phase. Between any two wires of a three-phase system there is a single phase. Each phase is an individual source of its own. All three phases are identical except that they are out of phase (Fig. 15-6).

It is possible to get three-phase service by using three separate transformers. Each transformer can be individually attached across two phase lines. The three separate outputs can then be used as a three-phase service source. This arrangement is called a transformer bank.

The secondaries of the transformers determine the type of service you get.

SINGLE PHASE

The schematic shows a transformer in the three-phase lines (Fig. 15-7). There is one primary winding for each phase and it is connected across two phase lines.

Typically, there would be 2400 volts in the primary winding. The primary coil acts as an inductive load across the 2400 volts.

The transformer could be a step down type. One-tenth of the 2400 volts, or 240 volts, would be developed in the secondary.

The secondary is isolated from the ground level of the primary. This is done to ensure that the only way the voltage can get from primary to secondary is by induction. The magnetic field passes right through the insulation of the windings and induces the secondary voltage.

This is important because the single-phase supply voltage is to be 120/240 (Fig. 15-7). The 240 volts are already floating in the secondary. All that is needed is to centertap the secondary winding and ground it. The centertap becomes neutral in the secondary. It is the same zero voltage level that is in the primary neutral, but at a different and more convenient place in the voltage level.

From the secondary centertap to either end is 120 volts. Each side is, in effect, a two-wire, 120-volt system. The total winding is a three-wire, 240-volt system.

This is one way that a residence or small building is supplied. The transformer must be large enough in wattage to carry the

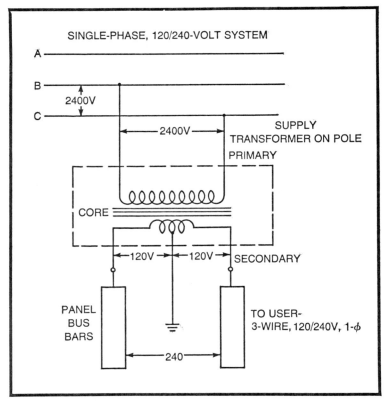

Fig. 15-7. A transformer in the three-phase lines with one primary winding for each phase connected across two phase lines.

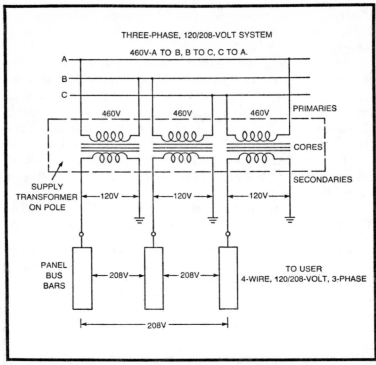

THREE-PHASE, 120/208-VOLT SYSTEM

460V-A TO B, B TO C, C TO A.

460V   460V   460V   PRIMARIES

CORES

SECONDARIES

SUPPLY
TRANSFORMER |← 120V →| ← 120V →| ← 120V →|
ON POLE

PANEL
BUS   ← 208V →   ← 208V →   TO USER
BARS   4-WIRE, 120/208-VOLT, 3-PHASE

|← 208V →|

Fig. 15-8. A 3-phase, 120/208-volt supply is taken from all three of the phase line conductors.

current requirements of the residence. For example, a residence with 200-amp service needs the following wattage from the power company's transformer:

$$P = EI = 240V \times 200A = 48,000W$$
$$P = 48 \text{ kW}$$

This is a nice sized transformer to be sitting in the pot on the pole.

FOUR-WIRE, THREE-PHASE, WYE

The schematics show three primary transformers attached across three-phase lines (Figs. 15-8 and 15-9). Each primary can pick up a separate 460 volts. The primaries are coupled to three secondaries.

The three windings do have a common neutral wire. If the three loads in the building are balanced, there will be no current in the

334

neutral. The three neutral currents, out of phase with each other by 120 degrees, will cancel the currents.

It is difficult to have three loads on all of the time, so there usually is some neutral current. However, the loads must be balanced as closely as possible.

The phase ends of the windings, though, have no connection with each other. They go to three separate phase lines. It is a delta-to-wye configuration.

The ratio of the primary to secondary is designed to induce 120 volts in each winding. This means that the 120 volts are dropped to zero across each winding to ground. Each winding is an individual two-wire, 120-volt type.

If you attach across any two windings using the neutral between them as ground, you have a three-wire system. At first glance, it

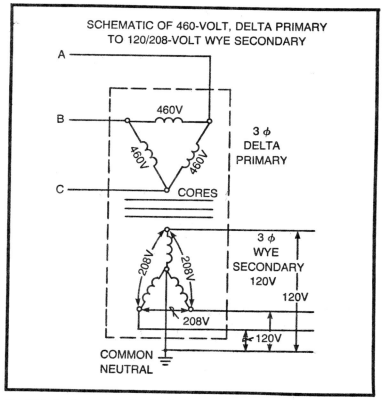

Fig. 15-9. Another way to view the 3-phase, 120/208-volt supply transformer is with this schematic.

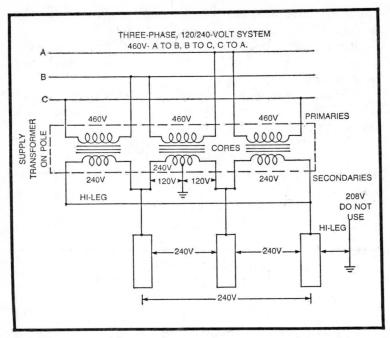

Fig. 15-10. The 3-phase, 120/240-volt supply is usually a delta-to-delta transformer.

looks like a three-wire, 120/240-volt system, but it is not. It's a three-wire, 120/208-volt system.

Even though both windings are dropping 120 volts to a common neutral, the two 120-volt pulsations are 120 degrees out of phase. Instead of adding up to 240 volts, they add to 208 volts. The 120-degree angle of impact (instead of being straight on) causes the 32-volt loss.

At any rate, the 120/208-volt, three-phase service is almost as good as the 120/240-volt service and is used commonly.

FOUR-WIRE, THREE-PHASE, DELTA

The schematic again shows three primary windings attached across three phase lines. It is exactly the same primary as the four-wire, 120/208-volt, three-phase. The three windings are in series. The primary is wired in delta.

The secondary is not wye. The three secondary windings are wired in series. They are in delta, too. Across each secondary winding 240 volts are developed (Fig. 15-10).

336

Note the middle winding. It has a centertap to ground. That winding is a three-wire, 120/240-volt, single-phase supply. Lots of buildings are supplied with service from one phase of a delta winding.

Across each winding there are 240 volts. Two windings do not have to be added together to get a voltage over 200 as with the wye type.

Only 120 volts are in each winding in the wye. Two windings are combined to get the 208 volts. In the delta there are 240 volts in each winding. The configuration of this transformer is called delta-to-delta.

Typically in a four-wire, three-phase, delta configuration, 120/240 volts are available from one phase and three 240-volt sources from all three phases. One winding is usually centertapped to ground the secondary and to provide 120 volts for small loads such as lighting.

When the delta is grounded with a centertap in one winding, you get what is called a "high leg." The high leg is to be marked and not used (Fig. 15-11).

The high leg is the connection in the secondary between the two windings not being centertapped. The high leg juncture is 208 volts to ground. The two windings not being centertapped are

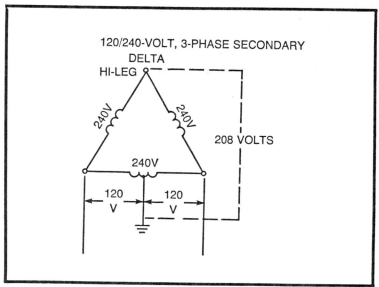

Fig. 15-11. Be careful of the high leg in a 3-phase, 120/240-volt secondary. Its odd voltage is best not used.

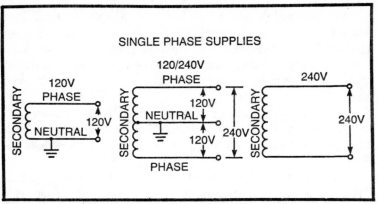

Fig. 15-12. Single-phase supplies can be obtained from a single-phase secondary in the utility company's transformer.

unavoidably added together, 120 degrees out of phase with each other, at the high leg point in the circuit.

The high leg is sometimes called the "wild leg." The important thing is to avoid using the junction to power anything. Your layout and balancing will be disturbed if you attach some form of load to the junction.

SINGLE-PHASE LOADS

Three-phase load problems often represent the difference between Journeyman's exams and Master Electrician's exams. The Journeyman is tested on single-phase loads and the Master is tested on both single-phase and three-phase loads.

One of the facts most difficult to learn is the difference between a single-phase load and a three-phase load. A single-phase load is simpler and lighter. Only one phase line is needed to power the load.

Typical single-phase loads are all sorts of lamps and appliances. These loads can be supplied with two-wire/120-volt, three-wire/240-volt, or two-wire/240-volt systems (Fig. 15-12).

The current, as stated, is 60 cycle and available in normal amounts. It has been shown that single-phase current is available from three-phase systems. The only major consideration in respect to the three-phase input is balancing. The single-phase loads on the three-phase input must be balanced carefully so that one phase doesn't deliver more current than the others.

In the 120/240-volt, single-phase service, truly only one phase is delivered to the load. In the 120/208-volt, single-phase service derived from the wye mode, the 120 volts are single-phase, but the 208 volts are actually composed of two phases. Two phase lines and two coils are involved (Fig. 15-13).

In effect, though, the two 120-volt phases add up to 208 volts, with a loss of 32 volts due to the 120-degree phase difference between the two inputs.

Calculating single-phase loads is not complicated because the current is coming from only one phase.

When calculating three-phase loads, remember that the current is threefold, since it is coming from three separate sources.

THREE-PHASE LOADS

A typical three-phase load is a three-phase motor. Instead of one winding, such as the one in a single-phase motor (Fig. 15-14), there are three windings in a three-phase motor (Fig. 15-15).

Instead of two hot lines, a single-phase motor has three hot lines.

Instead of two fuses and two switches, the single-phase hot lines have three fuses and a triple switch in the three-phase motor hot line.

A typical 5-hp, single-phase motor rated at 240 volts requires 28 amps of full load current, according to Table 43-148 of the Code book.

The conductors for the single-phase motor have to be able to carry the 28 amps, according to the motor regulations. Only one phase runs in and out and the current is 28 amps.

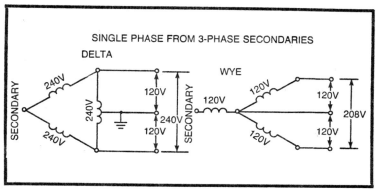

Fig. 15-13. Single-phase supplies can also be obtained from a 3-phase secondary in the utility company's transformer.

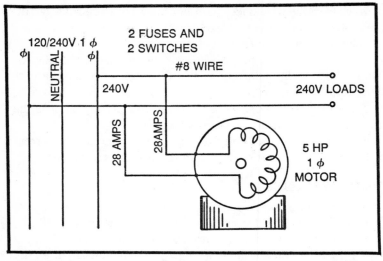

Fig. 15-14. A 5-hp, single-phase motor needs conductors and protection to handle 28 full load amperes. (According to Code the wire is rated 28A × 125%.)

Table 430-150 provides the 240-volt amp rating for a typical 5-hp, three-phase motor: 15.2 amps. That is roughly about half of the single-phase amp requirement.

That is because the 15.2 amps actually pertains to only one phase. But there are three phases and that means there are three

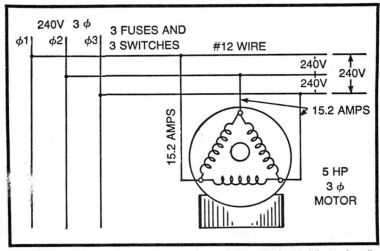

Fig. 15-15. A 5-hp, 3-phase motor needs conductors and protection to handle 15.2 full load amperes. (Code requires wire to be rated 15.2A × 125%.)

340

conductors, each carrying 15.2 amps. The capacity of the three wires together is 45.6 amps (Fig. 15-15).

Whereas the single-phase motor reqires a single conductor to handle 28 amps, the three-phase motor requires three conductors to each handle 15.2 amps.

The three-phase motor has three current inputs. It is as if you had three separate loads. A conductor is attached to each one of the three phases. This automatically balances the three loads of the three windings of the motor across the three phases.

When you calculate a three-phase service, you are to include each phase attachment. The single-phase loads are considered individually, and great care is taken to balance the loads by distributing the attachments over the three phases.

The three-phase loads of the three-phase service are automatically balanced, and each three-phase load has to be considered as three loads. Each of the three inputs in a three-phase motor or appliance draws its own current and consumes its own wattage.

Commercial Wiring

The wiring of residences and farms was discussed in Chapters 13 and 14. The lines of difference between the two were not great, and actual wiring techniques were similar. Once the hardware (the wire sizes, box types, etc.) is chosen, the equipment is installed with techniques that are universal.

It was shown that the key to a wiring job is calculating the size of the wires. After the feeder sizes have been computed, the hardware installation follows naturally. The same situation holds in commercial wiring. The service wires are calculated first and the rest of the job falls into place.

A typical commercial job could be the wiring of a small apartment house or a small store. Larger apartment houses and large shopping centers would be considered industrial jobs, as would factory installations. These will be discussed in Chapter 17.

In this chapter we will run through typical considerations and calculations for wiring up a small apartment house and a small retail store.

THE SMALL APARTMENT HOUSE

The small apartment house could have three, five, twenty, or perhaps as many as forty units. There is no fine line of demarcation between a small and a large apartment complex. For our example apartment house, we will cover the wiring calculations of a 36-unit, multiple occupancy building using 120/208-volt, three-phase service. This is all typical. The Code refers to apartment houses as "multiple occupancy" buildings and gives some rules about them in Article 230-72 (d). The Code's discussion of the distinctions between multiple and individual occupancies is confusing, but things will get clearer as this chapter unfolds.

Each apartment is considered to be an individual dwelling. Usually each apartment has its own meter, although in some cases the electrical service and maintenence are furnished by management through a superintendent. Since each apartment has its own service, there will be individual meters and individual disconnecting devices for each one.

Thus, each occupant will have access to his own main switch, panel, and overcurrent devices. In a one- or two-story building, there are two ways the service can be installed. One is to meter each apartment inside or immediately outside the apartment, treating each apartment as a separate home (Fig. 16-1).

The second way is to group and mark all the meters in one easily accessible space like a meter room or basement (Fig. 16-2). The metering of each individual apartment like a separate home is the most convenient way and this should be the first try. A meter room should be eliminated, if possible.

Unfortunately, the individual metering is limited to a one- or two-story building. If the building is three stories or higher, all the

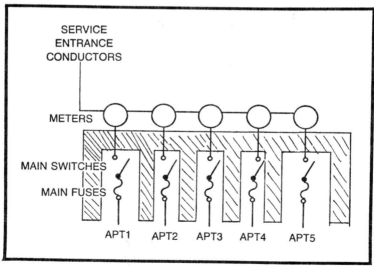

Fig. 16-1. In one- or two-story buildings each apartment can have its own outdoor meter, switch, and fuse.

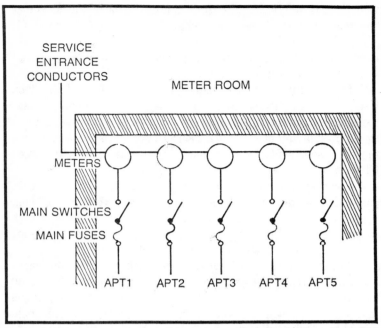

Fig. 16-2. A multiple occupancy building requires all of the service equipment to be installed in a common accessible space like a meter room.

meters must be grouped together where all tenants can get to them easily.

The exception to this is a building three stories or higher which has no tenants living above the second floor. This building can be treated as if it were a two-story building. Only the floors with living quarters are counted, not the non-living quarters floors.

When there are individual meters, the service calculations are identical to residential calculations and Chapter 14 covered that. In this case, each apartment can derive a 120/208-volt, single-phase service from the main 120/208-volt, three-phase service.

The 36 units of the apartment building we will examine are spread out over three stories. A grouping of all the individual services in an easily accessible place will have to be considered (Fig. 16-3).

This makes the 120/208-volt, three-phase service enter a *house service* panel first. Then the feeders from the house panel run to an individual-apartment panel. The important calculations are first to figure out the feeder sizes to each apartment and second to calculate the feeder size to the house panel. All of the panels are housed in the meter room.

THE SUPPLY

In this case, the 120/208-volt, three-phase supply is used. This is a four-wire, three-phase wye system. The secondary of the transformer on the pole is wound in a wye configuration. This is a very convenient type of service; its only drawback is that the voltage is 208 instead of 240. However, in most instances the 32-volt difference doesn't matter. The 208 volts will power appliances almost as well as the 240 volts (Fig. 16-4).

If you look at the circuit wiring diagram, you'll see that 120 volts pass from each point of the wye to ground. That provides three sources of 120 volts. These sources will be balanced among all of the apartments. Since there are 36 apartments, 12 apartments will be attached to each 120-volt point to satisfy the 120-volt, single-pole requirements.

There are 208 volts between any two points of the wye. This means that 12 apartments will be wired across each separate two points for the 208-volt, two-pole needs.

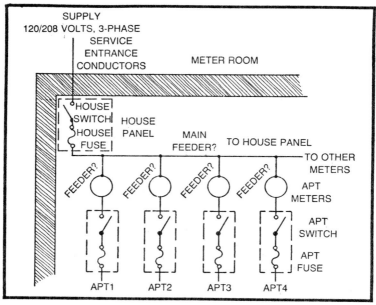

Fig. 16-3. The calculation in the text will enable you to determine the size of the main feeder and the apartment feeders.

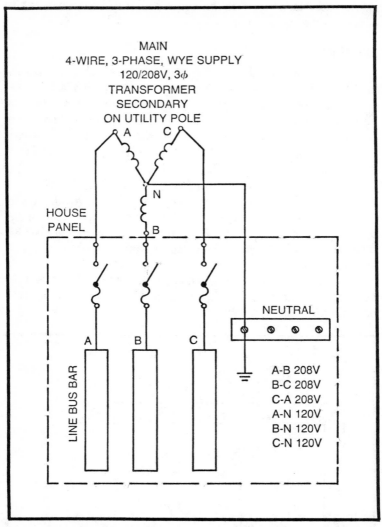

Fig. 16-4. The 120/208-volt, 3-phase supply can be used as three 120/208-volt, single-phase supplies.

The 120/208-volt, three-phase supply acts exactly like three 120/208, single-phase supplies.

THE LAYOUT

When a professional electrician is awarded a wiring job, the general contractor might give him the following requirements.

There are 36 apartments, 12 Type A and 24 Type B. Each apartment is to be served with a 120/208-volt, single-phase service.

In the A apartments, there are:

1800 square feet
1.0 kW dishwasher
8.0 kW range
3.0 kW water heater
9.0 kW air conditioner
10.0 kW heating strip

In the B apartments, there are:

2200 square feet
1.0 kW dishwasher
15 kW range
4.5 kW water heater
7½ hp air conditioner
9.0 kW heating strip

The electrician is ready to go, and it is assumed that he will install the minimum service needed. An apartment house of this nature is a money making proposition and wasting money on unneeded service equipment will not be appreciated by the investors.

It is better that a tenant occasionally have to reset a circuit breaker than to provide excess service that will probably never be needed.

The electrician looks over the requirements. He keeps in mind that Code book Article 220-16 recommends that each apartment have two 1.5-kilowatt small appliance circuits. There is no need for a laundry circuit of 1.5 kilowatts, since there is no laundry in the apartments.

He must first calculate the feeder sizes for the A and B apartments. Each will be 120/208-volt, single-phase service. Next he must size the main feeder. It is 120/208-volt, three-phase service.

THE A APARTMENTS

The total load of the A apartments is computed in the following manner according to Article 220 of the Code.

ARTICLE	USE	ϕ KW
220-2 (b)	Ltg. & Recept. 1800 × 3	5.4

ARTICLE	USE	φ KW	
220-16	Small Appliances 2—1.5 kW	3.0	
	Dishwasher	1.0	
220-19	Range		
Column C	8 kW × 80%	6.4	
	Water Heater	3.0	
	Total Unit Load	18.8	
Table 220-30	1st 10 kW at 100%	10.0	
	Remainder 8.8 at 40%	3.5	
	Air Conditioner	9.0	
Table 220-30	Heating Strip 10 kW at 65%	6.5	220-21
	Total A Load	22.5	

$$I = P/E = \frac{22.5}{.208} = 108\text{A Service Needs}$$

In Table 310-16, Note 3, the next highest service is 110 amps. THW copper #3 will take care of the phase feeder lines.

Table 250-94 recommends #8 copper as a good neutral. Therefore, each apartment will have feeders of two #3 wires and one #8 wire (Fig. 16-5).

THE B APARTMENTS

The total load of the B apartments is computed as the load of the A apartments was, except for a complication or two. According to Article 220 of the Code, the chart looks the same. As the electrician glances over the required loads, you'll notice that the air conditioner's load is expressed in horsepower. This will have to be converted to kilowatts to accommodate to the chart.

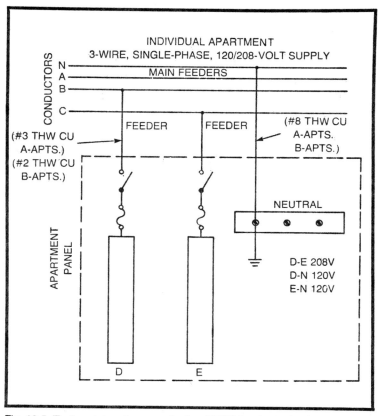

Fig. 16-5. Each apartment uses a 3-wire, 120/208-volt, single-phase supply.

Table 430-148 shows that the full-load current of a single-phase, 7½-hp motor is 44 amperes. (The 208 volts require a 10-percent addition to the 40 amps shown in the 230V column.)

$$W = E \times I$$
$$= 208V \times 44A$$
$$= 9152 \text{ watts or } 9.2 \text{ kW}$$

(Trick: use .208 when solving for kW. It saves converting watts to kilowatts.)

Now it's time to add up the total load of a B apartment.

ARTICLE	USE	φ KW
220-2 (b)	Ltg. & Recept.	6.6
	2200 × 3	

220-16	Small Appliances 2—1.5 kW	3.0
	Dishwasher	1.0
220-19 Note 1	Range 15 kW(8 + 1.2)	9.2
		4.5
	Water Heater	24.3 kW
	Total Unit Load	
Table 220-30	1st 10 kW at 100%	10.0
"	Remainder 14.3 at 40%	5.7
"	Air Conditioner	9.2
Table 220-30	Heating Strip 9 kW at 65%	~~5.85~~ 220-21
	Total B Load	24.9 kW

$$I = P/E = \frac{24.9}{.208} = 120A$$

Table 310-16, Note 3, indicates that the next highest service is 125 amps. THW copper #2 will take care of the phase feeder lines.

Table 250-94 tells that #8 copper will be a good neutral. Therefore, each B apartment will have two #2 feeders and a #8 neutral. These are the minimum sized feeders allowed by Code (Fig. 16-5).

THE HOUSE SERVICE

While there are some other loads in an apartment house, like outside and hallway lighting, a special laundry room and the like, they are going to be ignored in this example. The house service will be calculated as if the apartments were the only loads drawing current from the main feeders and panel.

The size of the main feeder is the next important consideration. All of the 36 apartments are going to draw their loads collectively through the main feeders. The main switch and main overcurrent protection can be figured easily after the feeder sizes are determined.

The main feeder is a four-wire type with three phase lines and a neutral (Fig. 16-4) The house load in some instances might have to supply three-phase motors, but in this case the three phases distributed to the 36 apartments are operating only single-phase equipment. There is plenty of two-pole (208-volt) equipment, but it is operating on one phase.

The three-phase service never reaches an apartment panel. The apartment feeders are only three-wire, two-phase lines and a neutral. Between each phase line and neutral are 120 volts. Between the two phase lines are 208 volts (Fig. 16-5).

The four wires from the utility pole attach to the house panel. The three phase lines (A, B, and C) attach to the three line bus bars in the panel. The fourth wire, N, neutral, attaches to the neutral bus bar in the panel.

Each apartment in turn takes current from two of the hot bus bars and from neutral. To balance out the apartments, the A's and B's are distributed in the following fashion.

Four A's and eight B's are attached to two phases and neutral. Since there are three possible hookups (as shown in the illustration), each group of 12 apartments is balanced around the wye input (Fig. 16-6).

The main service feeds all 36 apartments and the individual services feed each individual apartment. The A apartments use 110-amp service and the B apartments use 125-amp service.

CALCULATING THE MAIN FEEDERS

Calculating the main feeder is tricky, but it is not overly difficult. The idea is to add together all the loads and apply the proper demand factor.

The demand factor is extracted from Table 220-32 of the Optional Calculations. The table is straightforward but quite astounding. If we look at the demand factor percentages, they range downward from 45 to 23. This means, for example, that if the total house current load were computed at 2500 amps, in this 36-unit apartment house we could demand the feeder by 30 percent. 2500 × 30% = 750 amps. We'd only need a service to handle 750

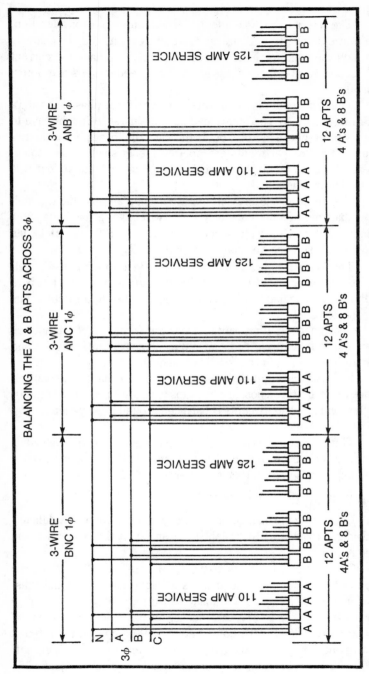

Fig. 16-6. The 36 apartments must be balanced carefully, since they are attached to the mains.

amps (actually, we'd probably go to an 800-amp service) for a total 2500-amp load!

The Code writers feel that an apartment house with no vacancies, everyone home at dinner on a hot summer night, everyone cooking full blast and keeping cool with the air conditioner is a remote possibility. It is safe, however, because the main overcurrent device would blow if the above situation should come to pass. There would be some inconvenience, but little danger. In the meanwhile, the compensation will be practical savings in installation costs.

To calculate the main feeder, we will divide the total load into four parts:

1. The total *UNDEMANDED* load, minus the range load, of the A apartments.
2. The total *UNDEMANDED* load, minus the range load, of the B apartments.
3. The total range load for all apartments, according to Table 220-19.
4. The total heat and air conditioning loads, using 220-21 (Non-Coincident Loads). In other words, the total heat of air conditioning for A and B apartments, using whichever one is larger in each apartment. It could be air and air, air and heat, or heat and heat, however it works out.

Adding these four numbers together gives us the total house load. Then, using Table 220-32, we demand that load. Kilowatts are converted to amps by dividing by .360 (since it is a three-phase system). $.208 \times \sqrt{3} = .360$.

Once the amps are obtained, the service sizes can be routinely found in the Code book.

TOTAL UNDEMANDED LOAD MINUS RANGES

The 36 apartments have total unit loads of 18.8 kilowatts for the A's and 24.3 kilowatts for the B's. This does not include the air conditioners or heaters, since they are calculated separately in Table 220-30. We should deduct 6.4 kilowatts for the range demanded load of the A apartments.

$$\begin{array}{r} 18.8 \text{ kW} \\ - \underline{6.4 \text{ kW}} \\ 12.4 \text{ kW} \end{array}$$

The 12.4 kilowatts represent the total undemanded load of one A apartment. There are 12 such apartments. $12 \times 12.4 = 148.8$ kW.

That means that the A apartments will draw 148.8 kilowatts without the ranges.

The total undemanded B load is 24.3 kilowatts before deduction of the range demanded load.

$$\begin{array}{r} 24.3 \text{ kW} \\ - \ \underline{9.2 \text{ kW}} \\ 15.1 \text{ kW} \end{array}$$

Since there are 24 B apartments, $24 \times 15.1 = 362.4$ kW.

A Apartments	148.8 kW
B Apartments	362.4 kW
Total Undemanded Load (Minus ranges)	511.2 kW

THE RANGE LOAD

Computing the house range load is the difficult part of the main feeder calculations. You should follow this section and keep reviewing it until you understand.

First of all, there is the problem of 36 ranges being supplied by three separate phases. There are three hot lines and a neutral. All of the ranges get attached to neutral. There is no confusion there.

However, there are only two hot lines on a 208-volt range and there are three hot lines in the main service. This means that any one range is attached to only two of the three hot wires. Between any two hot service lines there are 208 volts in single phase.

Count the number of ranges on any two hot lines. Only two-thirds of the ranges, in this case 24, are on any two hot lines.

Furthermore, there are two kinds of ranges, 8-kilowatt and 15-kilowatt, balanced on the hot lines. There are 12 8-kilowatt and 24 15-kilowatt ranges. Two-thirds of 12 and 24 are 8 and 16.

These are the single-phase range loads of the individual apartments. 8×8 kW = 64 kW, 16×15 kW = 240 kW. The three-phase current distributed to the apartments as single-phase causes this confusion. The three-phase input powers the single-phase ranges in this manner.

To further complicate the matter, the ranges in the preceding apartment feeder calculations were demanded according to Table 220-19 on an individual basis. For this main feeder calculation, the ranges are not demanded on an individual basis, but on a total basis. Table 220-19 is used again, but in an entirely different way. The ranges are demanded in total, not individually.

Another confusing element is the way the ranges are deducted from the unit loads we discussed in the previous section.

If you understood the preceding at the first reading, you are a genius. It is confusing and you must pick your way through, over and over, until you finally understand it.

At any rate, since there are 12 single-phase ranges in the A apartments, they load the main feeder like 8 ranges, since the main is three-phase. Also, there are 24 single-phase ranges in the B apartments loading the main feeder like 16 ranges.

Referring to Table 220-19, 8 ranges at 8 kilowatts each totals 64 kilowatts and can be demanded by 36 percent according to Column C. 64 kW × 36% = 23 kW. 12 ranges at 15 kilowatts each totals 240 kilowatts and can be demanded at 31 kilowatts plus 15 percent of the 31 kilowatts. This is obtained from Column A and Note 1. 31 kW × 15% = 4.65 kW.

$$\begin{array}{r} 31.00 \text{ kW} \\ \underline{4.65 \text{ kW}} \\ 35.65 \text{ kW} \end{array}$$

Our two demanded range loads for the main feeder calculations are 23 kilowatts for the A apartments and 35.65 kilowatts for the B apartments.

$$\begin{array}{r} 23.00 \text{ kW} \\ \underline{35.65 \text{ kW}} \\ \text{Total Range Load} \quad 58.65 \text{ kW} \end{array}$$

HEAT AND AIR CONDITIONING LOADS

The heat and air conditioners are typically wired together on the same thermostat, even though they have different interior wiring distributions. It is not possible to turn them both on at the same time. Therefore, one or the other can be on, but never both.

To compute the main feeder capacity needed to power them, the one with the larger load is considered. In both A and B apartments, the air conditioners consume more kilowatts than the demanded heat strips. Therefore, the kilowattage of the air conditioners is used for the computation.

There are 12 air conditioners at 9 kilowatts and 24 at 9.2 kilowatts. (We calculated this previously.)

$$12 \times 9 \text{ kW} = 108 \text{ kW for the A's}$$
$$24 \times 9.2 \text{ kW} = 220.8 \text{ kW for the B's}$$

$$\begin{array}{r} 108 \text{ kW} \\ \underline{220 \text{ kW}} \\ 328 \text{ kW} \end{array} \quad \text{Total Air Conditioner Load}$$

GRAND TOTAL LOAD

The four total loads are

Total Undemanded Load—A Apts. (minus ranges)	148.80 kW
Total Undemanded Load—B Apts. (minus ranges)	362.40 kW
Total Range Load—A & B Apts.	58.65 kW
Total A/C Load—A & B Apts.	328.00 kW
Grand Total	897.85 kW

DEMANDING WITH TABLE 220-32

Once the grand total is arrived at, we turn to Table 220-32. To use the table we must have a building that conforms to all of the table's requirements (a and b). If all of the requirements are met, then we can look up the demand factor. In this 36-unit apartment house, the factor is 30 percent.

We multiply 897.85 kilowatts by 30 percent and determine that the feeder must provide a minimum of 269.4 kilowatts. Then we convert kilowatts to amperes.

$$\frac{269.4 \text{ kW}}{.360 \text{ volts}} = 748 \text{ amperes}$$

An 800-amp service will do nicely.

Why use 360 volts as the divider? Simply because there are 120 volts on each phase to ground, and there are three phases. The product is 360 volts.

Actually, the 360-volt figure is the result of 208 being multiplied by the square root of 3. The mathematical explanation is on an engineering level and beyond the scope of this book. No need for you to worry about it. Even most Master Electrician's exams don't ask for that kind of explanation. However, study this example carefully. It's very instructive.

SMALL STORE

The task of wiring a small retail establishment (with floor space of between 500 and 2000 square feet) is a normal one for the average electrician. Most small stores require power for lighting, receptacles, heating, air conditioning, and a small water heater.

Power is typically 115/230-volt, single-phase or 120/208-volt, single-phase. The service can vary between 75 and 200 amps. The calculations are usually easy. The loads are not demanded and equipment nameplate ratings serve to indicate how to wire the place.

Table 220-2 of the Code book lists the general lighting loads. As you can see, they vary from ½ watt per square foot in a garage to 5 watts per square foot in an office. Ordinarily, small stores are wired at approximately 3 watts per square foot.

Because store lighting is usually considered "continuous" burning, the individual circuits are calculated at 80 percent of rated output. For example, a 15-amp lighting circuit is not permitted to furnish more than 15 × 80% = 12 amps, which translates into 1380 watts.

Code book Article 220-12, Show-Window Lighting, recommends that 200 watts per linear foot be installed. If we consider a store (Fig. 16-7) with 1800 square feet of floor space (30' by 60'), we have the following loads to be wired.

Lighting & Receptacles—1800 sq. ft. × 3 watts/sq. ft. = 5400W = 5.4 kW. If each 15-amp circuit provides 1380 watts, then 4 circuits (5400/1380) are needed. The circuits will use #14 wire.

Show Window—According to the sketch, each of the two show windows is 14 feet long (measured linearly horizontal along the base). 28 lin. ft. × 200 watts/lin. ft. = 5600W = 5.6 kW. If the inspector considers the show window to be an instance of continuous burning, then the wattage will have to be increased to 125 percent. 5.6 kW × 1.25 = 7 kW.

If we use 20-amp circuits here and they are 2300 watts each, then 7 kW/2.3 kW = 3 circuits needed. These circuits can not only handle the show window but the doorway entrance lighting, too. They'll require #12 wire.

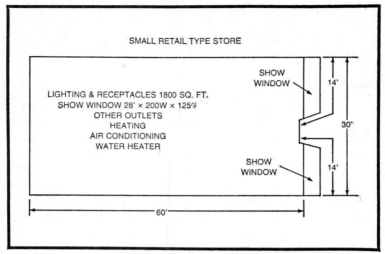

Fig. 16-7. A small retail store is usually easy to calculate.

Other Outlets—The store has outlets other than the 15-amp type. It has the 20-amp, 2300-watt type, too. For a building of this size (and for most small stores and service establishments), three circuits of 2300 watts each is sufficient. There is a maximum load rating on each outlet of 1½ amps, which at 230 volts is 1½ × 230 = 345 watts.

The number of outlets to be installed on each circuit is determined by dividing 2300 by 345. The answer is 6.6 outlets. The total of the three circuits would be about 20 outlets. They would be connected with #12 wire.

Some of these outlets would be mounted above the show windows so that lighted signs and displays could be plugged in easily.

TOTAL LOAD

Assuming that the heating is 7 kilowatts, the air conditioning 6 kilowatts, and the small water heater 1 kilowatt, the total load is computed in the following way. Notice that in commercial calculations, both heating and air conditioning are counted. Non-Coincident loads (Article 220-21) are usually ignored.

Article	Use	ϕ kW
Table	Ltg. & Receptacles	
220-2 (b)	1800 × 3	5.4 kW

358

Show Window	5.6 kW
200 × 28	
3 Recpt. Ckts	6.9 kW
Heating	7.0 kW
Air conditioning	6.0 kW
Water Heater	<u>1.0 kW</u>
	31.9 kW

We can figure I = P/E = 31.9/.230 = 138.7 amps. Thus 150-amp service can be used. The load can be conducted nicely with #1/0 THW copper. As you can see, calculating a small store is quite easy when compared to calculating a multiple-dwelling apartment house or even a residence.

Introduction To Industrial Wiring

Perhaps a good line of demarcation between industrial wiring and residential-rural-commercial wiring can be drawn after looking at the distribution systems for each.

INDUSTRIAL DISTRIBUTION SYSTEMS

Industrial systems use mammoth amounts of electrical energy in comparison to the loads of residential-small commercial wiring. The distribution systems reflect this fact. Typical small systems are 2-, 3- and 4-wire, single- and three-phase types delivering voltages around 120, 208, and 240. The large systems use 3- and 4-wire, three-phase circuits delivering voltages like 277, 480, 2400, 4160, 4800, 7200, 13,800 and more.

The commonest of the higher voltage options offered to industrial users is the 277/480-volt, three-phase, wye configuration (Fig. 17-1) and resembles the 120/208-volt, three-phase in operation. From any point in the wye to ground there are 277 volts. Between two points there are 480 volts. A separate phase comes off each of the three 120-degree points (Fig. 17-2).

In office and commercial type buildings, the 277/480-volt service is wired to each individual floor. The 480-volt, three-phase is then connected to a floor panel and to the motors on that floor.

Between each phase, the 277-volt mode is sent to the heavy lighting loads. The lights are built with 277-volt ballast transformers and are powered with 277 volts.

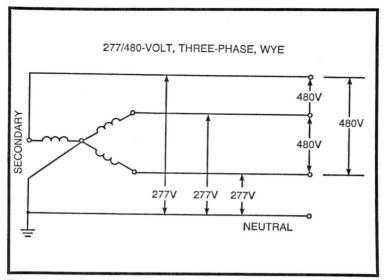

Fig. 17-1. A typical 277/480-volt, three-phase, wye industrial service.

Step down transformers convert the 277/480 volts to 120/208 volts. These lower voltages can then power the conventional lighting, receptacle and appliance loads.

Where the 277/480 voltage is flowing (which is between floors to motors and ballasts of fluorescent lights, etc.), the copper wire sizes are small because the current flow is relatively small. The 277/480 voltage flows in the long lengths of copper. The 120/208-volt lines are designed to be as short as possible. As long as the number of 120/208-volt loads is kept small, the cost of the step down transformers is offset by the large savings in copper and fittings to handle the smaller current flow from the larger voltages.

TYPICAL 277/480-VOLT, THREE-PHASE LOAD

When you look up at a tall office building or over a short building which covers a full city block, the vertical or horizontal structure is vast. The electrical needs are awesome. A huge quantity of energy is being used, but if the rooms are considered one by one, the wiring requirements are not too complex to calculate or to install.

For example, take a typical office in a large building. It receives a 277/480-volt, three-phase service. It is a large office, 150′ by 50′ with 7500 square feet of floor space. There is a computer area in the right top corner, overhead banks of fluorescent fittings, smaller

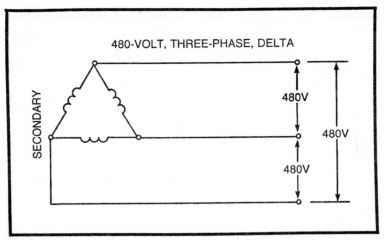

Fig. 17-2. A 480-volt, three-phase, delta service is often used in factories.

individual desk lighting, receptacles (spaced liberally), and air conditioning with reverse cycle heat (Fig. 17-3).

An electric supply room is built in the left top corner of the office space. The electric room has a bus duct riser containing 480-volt, 4-wire, three-phase power. The size of the bus duct feeder is calculated on the basis of the total office load.

The bus duct attaches to a large distribution panel in the electric room. The main panel distributes the energy to six load centers: computer (two), air conditioning equipment (one), and panels (three). There are two 277/480-volt panels and one 120/208-volt panel. Most of the loads need 277/480-volt service, while some desk lighting and receptacles need the lower 120/208-volt service. A transformer in the electric room steps down the 277/480 voltage to 120/208 voltage for the one lower voltage panel.

The transformer is called a delta-wye, since its primary is wound in a delta configuration and its secondary in a wye. The wye output allows three single-phase, 120 or 208 voltages to be extracted from the three-phase, 480-volt input.

The fluorescent overhead fixtures which represent most of the lighting are energized through 277-volt ballast transformers. The two 277/480-volt panels can take care of all the lighting. Each 277-volt ballast can be connected between one of the phases and neutral, just like the 120-volt ballast can. The 277/480-volt supply is exactly like the 120/208-volt, 4-wire supply, except for the higher voltage values.

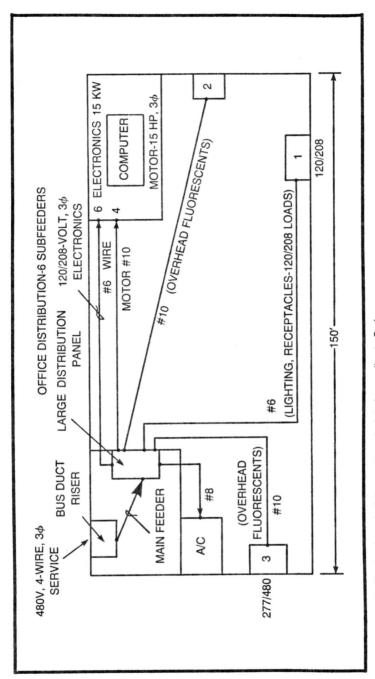

Fig. 17-3. A small office in a large building can be calculated according to Code.

363

The computer's small lights and tiny motors operate on 120/208-volt, single-phase current and its electronics operate on 120/208-volt, three-phase current. The 120/208-volt, single-phase current can be pulled from the small panel. The larger motor loads and electronics are supplied by feeders from the electric room main panel.

The air conditioner, which is at least a 20-hp size, is served well with 480-volt, three-phase current, also supplied by the main panel. The air conditioner according to Table 430-150 draws 52 amps, which is $E \times I = .480 \times 52 = 25$ kW.

CALCULATING THE LIGHTING LOAD

The 277/480-volt supply is no different than the 120/208-volt, three-phase supply, except for the larger voltages. The larger voltages provide savings on copper conductors. The higher the voltage, the less the current which must be dragged through a conductor to power a load..

Since the floor space of the building totals 7500 square feet and the Code in Table 220-2 (b) specifies a minimum unit of 5 watts per square foot, then $7500 \times 5 = 37.5$ kW.

Table 220-11 lists no demand factors, so the 37.5-kilowatt figure without adjustment represents the lighting load. All the lighting will be of the 277-volt type, except for the electric room which receives ordinary 120-volt current (which is not counted here). The electric room lighting is considered as part of the receptacle load at the 120/208 voltage.

A #12 conductor with 20-amp protection is used. The office lighting is considered continuous burning, so the circuits can only be loaded to the 80-percent point. Each lighting circuit can then be loaded to 80 percent of 20 amps, which is 16 amps. Since the bulbs are all of the 120-volt type, $16A \times .120 = 1.92$ kW. That means that each circuit can only provide 1.92 kilowatts.

The total of the 277-volt circuits will be the total wattage (37.5 kW) divided by 1.92 kW. This amounts to 19.5 circuits. Since it is a three-phase board, the circuits have to be balanced over three lines. The next higher number divisible by 3 is 21.

However, there are two 277/480-volt panels. Each panel has to accept half of the circuits; yet the number of the circuits must be divisible by 3. So the number 24 is substituted for 21. Therefore, each panel gets 12 circuits. These can be balanced and each circuit is loaded less than 80 percent.

CALCULATING THE 120/208-VOLT LOAD

The 120/208-volt panel takes care of the desk lighting, the receptacles, and the overhead light in the electric room. #12 wire is used to run 20-amp circuits which can be loaded to 100 percent capacity (2.4 kilowatts per circuit).

Each receptacle is rated at 1.5 amps. In an office of this size, about 80 assorted receptacles could be installed. At 1.5 amps each, and providing 120 volts each, 1.5A × 80 × 120V = 14,400 watts or 14.4 kilowatts.

If 14.4 kilowatts is total and each circuit consists of the 2.4 kilowatts, then exactly 6 circuits are needed for receptacles and assorted lighting. A couple of spare circuits should be installed in the panel in case some future needs arise. This brings the total of 2.4-kilowatt circuits to 8.

COMPUTER MOTOR

The computer's three-phase motor is a 15-hp model. According to Table 430-150, the motor draws 20 amps of full load current at 480 volts. This three-phase voltage is pulled from the electric room on its own cable rather than from any of the panels.

THE SUBFEEDERS

The subfeeders are now easy to calculate. There are six of them. They are,

1. 120/208-Volt Panel— The size of the feeder from the main panel to the 120/208-volt panel is determined by the total panel load (the 19.2-kilowatt receptacles and the 200-watt electric room-loads). This non-continuous load of 19.4-kilowatt, three-phase line current is found by dividing the watts by the phase voltage times the square root of three. The phase voltage is 208 × $\sqrt{3}$ = 360. To convert kilowatts to amps, .360 is used.

$$I = kilowatts/volts = \frac{19.4}{.360} = 54 \text{ amps}$$

Table 310-16 shows that #6 TW wire is able to handle 55 amps.

2. & 3. 277/308-Volt Panels— These identical panels with 12

circuits each supply continuous burning loads. The total load is 1.92 kW × 24 = 46 kW. Each panel supplies half of the load. ½ of 46 kilowatts = 23 kilowatts.

$$\text{Total, One Panel} = \frac{23,000}{480 \times \sqrt{3}}$$

$$= \frac{23,000}{831.4} = 27.66 \text{ amps}$$

A #20 TW will handle the 27.66 amps. The voltage drop across the expanse of the office is not a consideration because three #10 TW's will feed each panel.

4. 15-HP Computer Motor—The 15-hp, three-phase motor has a full load current of 20 amps. The motor amps have to be multiplied by 125 percent to conform with Code. 20 × 1.25 = 25 amps. A #10 wire can carry the 25 amps in the three-phase lines.

5. 20-HP Air Conditioner— The air conditioner motor in Table 430-150 has a value of 27 amps on 480-volt, three-phase current. Again the Code requires a wire at 125 percent of the load. The load is multiplied: 27A × 1.25 = 33.75 amps. The nearest size of wire in Table 310-16 is #8 TW.

6. 120/208-Volt Supply to Computer— The computer needs its own three-phase, 120/208-volt system. This is taken from delta-wye step down transformers. The nameplate rating of the computer system is 15 kilowatts. According to Code, the subfeeder can be sized at 100 percent of the load.

In the three-phase formula, 15 kilowatts is divided by 208 $\sqrt{3}$ = 360.

$$I = \frac{15 \text{ kW}}{.360} = 42 \text{ amps}$$

The next highest wire size is a #6 TW.

CALCULATING MAIN FEEDER

Between the bus duct riser and the main electric room panel, there is a main feeder. Bus duct comes with a nameplate stating its ampacity rating.

The main feeder though must be calculated and chosen. All the power that is drawn at 480 volts is added to the power at 208 volts. It is best to convert the currents (if their values are known) into kilowatts and then add them together.

The 15-hp computer motor draws 20 amps at 480 volts. $P = E \times I$, but with three-phase power, the resulting P needs to be multiplied by the square root of three.

$$P_{3\phi} = E \times I \times \sqrt{3} = 480v \times 20 \text{ A} \times \sqrt{3}$$

$$= 16,628W = 16.6 \text{ kW}$$

The 20-hp air conditoner motor is calculated in the same way.

$$P_{3\phi} = E \times I \times \sqrt{3} = 480V \times 27A \times \sqrt{3}$$
$$= 22,447W = 22.4 \text{ kW}$$

A complication arises since there are two motors on the main feeder. The Code in Article 430-24 requires the feeder to carry 125 percent of the larger motor plus the other one.

The larger is the air conditioner motor. 22.4 kW × 1.25 = 28 kW. The total feeder capacity is derived by adding the 16.6 kW, for an answer of 44.6 kW.

Now the 480's and 208's are added.

15 hp and 20 hp	
Total Motors at 480	
120/208 Panel	44.6 kW
8—2.4 kW Circuits	19.2 kW
120—208 to Computer	
Nameplate	15.0 kW
2—277/480 Panels	
Lighting	46.0 kW
	124.8 kW

The current in the phase lines can be calculated with

$$I = \frac{\text{total kW}}{.480 \times \sqrt{3}} = \frac{124.8 \text{ kW}}{.830V}$$

$$= 150 \text{ amps}$$

The service amperage is reasonably low, even with the high kilowattage required. This is because the 277/480 voltage is used instead of a lower voltage.

ADEQUATE INDUSTRIAL WIRING

In residential wiring and even small commercial wiring, the homeowner or small business man has a personal stake in the wiring system. Since he is so involved, extra capacity is usually built into the individually designed systems. A larger service that can handle future needs, extra receptacles for added convenience, and all kinds of personal desire electrical appliances are installed. Not too much expense is spared in swimming pools, sauna baths and more than adequate air conditioning. Saving on expenses isn't a paramount consideration in these personalized electrical systems.

In large buildings, plants and shopping centers, the electrical system's cost is part of an investment. A factory president sitting a thousand miles away, and with no personal stake in convenience, is concerned solely with installing a safe, adequate system for the smallest cost possible. Except for those companies which are so rich that money is not a major consideration, the lowest reliable bid will get the contract.

Accordingly, an electrical system that is safe will be installed. However, after several years pass and all of the factors have been evaluated, the system might turn out to have a lot of problems. What seemed to be the most economical system when it was installed may turn out to be the least economical in the long run.

To install an excellent industrial wiring job, a lot of factors should be evaluated during design. The major design considerations have evolved over the years. Great thought should be given to six considerations: 1. Initial Investment; 2. Flexibility; 3. Efficiency; 4. Voltage Regulation; 5. Service Continuity; 6. Eventual Annual Cost of Operation and Maintenance.

A LOOK AT THE DESIGN CONSIDERATIONS

1. Initial Investment. It turns out that the minimum service permitted by Code is not usually the best value in the long run. While the installation of the least expensive wire saves initial money, a lot of money is wasted in maintenance because overcurrent devices open a lot more often. The same type of operational extra expense occurs in every aspect of the system.

2. Flexibility. A system that has little flexibility is O.K. for the original uses. However, as time goes by, changes occur. Different levels of skill, changes in technology, even changes of tenants in large buildings, require major changes in electrical systems. If the system does not have flexibility built in, it might have to be completely replaced, which is a major expense.

3. Efficiency. If the efficiency of a system is not good because corners were cut on expenses, large losses continue as the system is used.

4. Voltage Regulation. Enough capacity should be built in so that no load can seriously cause the voltage to change. The ideal system will maintain its operating voltage continuously no matter what load suddenly turns on.

5. Service Continuity. This is built into a system and is called anti-fault. This means good insulation, carefully controlled ambient temperature conditions, etc. The ideal system would never develop a fault and would maintain continuous service.

6. Operation and Maintenance. During design a lot of thought goes into operation and maintenance. The ideal system never breaks down and can be serviced periodically with easy inspection, appropriate tests, and various maintenance procedures. Of course, no system is perfect, but that is the goal to strive for.

TYPES OF INDUSTRIAL SYSTEMS

Early in this chapter we discussed the laying out of a good sized office. No matter what size or shape a building takes, each individual office, store, or manufacturing area is calculated in this manner. The Code directs that a safe system be installed.

However, the distribution part of the system—the part that conveys the electricity from the street to the loads—breaks down into five classifications.

1- Groups of small buildings
2- Industrial plants
3- Vertical buildings up to about 12 stories
4- High rise buildings
5- Large horizontal buildings like shopping centers or one-story office buildings that occupy a large area.

Groups of small buildings are typically schools, hospitals, and the like. Each building receives its own service and the wiring is not much different from that of residences and shops.

Industrial plants and their wiring are systems unto themselves. Ordinarily, an industrial plant has its own distribution system. The system is designed to meet the plant's special needs. The utility company brings in a very high kilovoltage to the plant's electrical substation. The plant takes it from there. An extremely large industrial plant may even generate its own electricity.

Vertical buildings like office buildings, stores, apartments, hotels, and hospitals use what is called a "network system."

Tall buildings are usually built because there is very little ground available in the area. Tall buildings are almost always built in the densely loaded centers of cities. The buildings are wholly dependent on elevators, fire protection, water supply, heating, and ventilation. The electrical systems must be reliable. Reliability is probably the most important factor.

THE SIMPLE SPOT NETWORK SYSTEM

The so called network system is best known for its high reliability. No single fault in any part of the system will shut down the entire system. A fault will only prevent energy from being delivered to a small area. Most common faults are automatically cleared without interrupting service to any of the loads, including those loads in the area where the fault occurred.

Network systems also have built in flexibility. New loads, new tenant needs, and so on are easily accommodated by the existing system. There is excellent voltage regulation and high efficiency. What are the features that enable network systems to do all this?

The idea is to install a system with considerable backup capability. Figure 17-4 shows an example of a power supply feeding into three primary feeding breakers. The three currents then feed through the primary feeder to the various banks of transformers.

Each feeder is attached to a transformer. The transformers are arranged in banks of three. The outputs of the transformers pass through breakers and then are attached to bus bars. Each bank of transformers is wired to the bus bar in parallel. If a fault occurs in one of the feeders, one of the feeder breakers opens and one of the transformers receives no current.

The other two transformers in the bank, however, continue to receive current, and they are built with plenty of extra capacity. The loads are still supplied with current from the two remaining transformers. Perhaps the lights will blink when a fault occurs in a feeder, but that is about all.

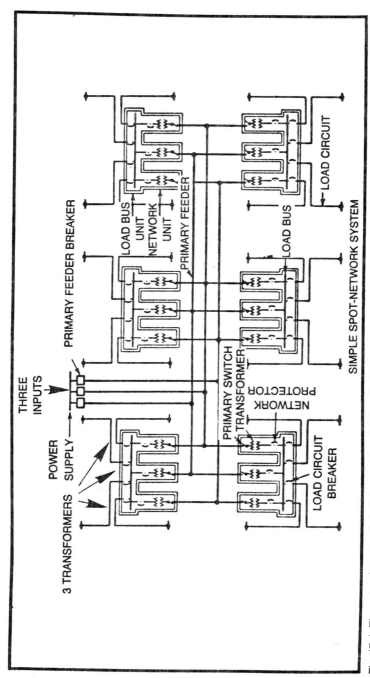

Fig. 17-4. The network system is built to provide reliability. If a fault occurs, the system continues to operate.

371

Work can then be performed on the dead feeder. The fault is cleared and the feeder breaker reset. The elevators have continued to run all the while. The air conditioners have kept everyone cool and the fire protection and escape routes have remained in force. Probably no one but the electrician knew that a fault had occurred.

The system is reliable. The extra capacity of the backup transformers provides the plus items of good voltage regulation and efficiency.

LARGE HORIZONTAL BUILDINGS

The commonest type of large horizontal structure is the shopping center. All three classes of centers, neighborhood (5000–20,000 people), district (20,000–100,000 people), and regional (over 100,000 people), are horizontal. Another common type of large horizontal structure is the one-floor office complex, erected where land costs are not a problem.

The physical layout of the shopping center or office building is often influenced by the nature of the electrical system used. A typical type of electrical system used in horizontal buildings is the *radial* system.

A simple radial system uses a single substation. The utility company supplies current at a high supply voltage. The voltage is then stepped down to the conventional levels of 120/208, 277/480, and so forth.

Low voltage feeders are run from the substation to panel locations that are set up to service particular loads (Fig. 17-5).

The operation of this system is quite like the operation of the system whereby the utility company provides current to a large number of residences. The radial system can take advantage of the diversity of the loads and the transformer can be demanded to a smaller size than the entire load would otherwise require.

The simple radial system has all of the disadvantages of the residential type of service. Voltage regulation and efficiency, all on low voltage feeders and one source, without backups are low. A fault in the substation transformer circuit could shut down the whole shopping center. Furthermore, it might not be possible to restore power to the circuit until the fault has been cleared and the repairs made. This could be very inconvenient.

However, the safety factor is not too serious because a one-floor horizontal building is easily evacuated. In a skyscraper, this sort of installation could not be tolerated.

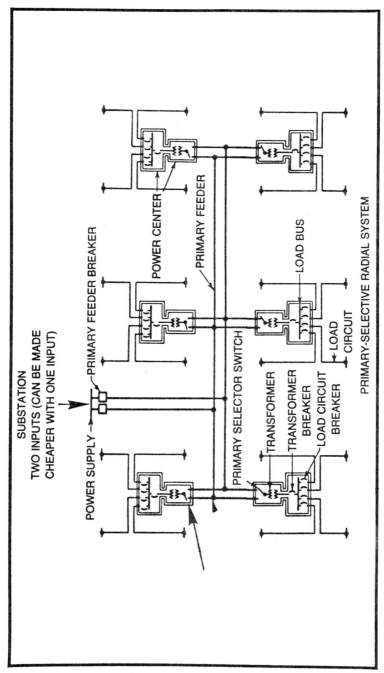

SUBSTATION
TWO INPUTS (CAN BE MADE
CHEAPER WITH ONE INPUT)

POWER SUPPLY

PRIMARY FEEDER BREAKER

POWER CENTER

PRIMARY FEEDER

LOAD BUS

LOAD
CIRCUIT

PRIMARY SELECTOR SWITCH

TRANSFORMER

TRANSFORMER
BREAKER

LOAD CIRCUIT
BREAKER

PRIMARY-SELECTIVE RADIAL SYSTEM

Fig. 17-5. The simple radial system resembles the way the utility company delivers power. It has a substation.

The simple radial system is much cheaper to install than the network system because the backup of extra transformer and wiring is not needed.

The system can be made more flexible by adding more primary feeder circuits, but this costs extra money and is difficult to justify in an investment situation. However, if a fault in a primary feeder causes the primary feeder breaker to open, the following can be done.

The manual breaker for the faulted feeder can be opened. This relieves the short circuit and all the other primary feeders can be turned back on by resetting the primary feeder breaker.

Only the faulted circuit will not be energized. It must wait until the fault is cleared.

The radial system and the network system have a large number of variations according to their specific needs and actual applications.

ALLOWING FOR THE FUTURE OF FEEDERS

When an industrial type of project is designed and growth planning is considered, a rule of thumb is to size the feeders to allow at least a 50-percent increase in load. Even this is usually not enough as years go by.

This is quite different from residential wiring design. Lots of economy homes are built with no future growth planned. The demand factors are taken to the limit and the thinnest feeder practical is installed.

In horizontal building layouts, the problem is not serious. Additional feeders can be installed alongside existing feeders. The radial type of supply lends itself to such a modification.

In vertical buildings, though, the problem is extreme. Once a skyscraper is installed with a network type of supply, the users are virtually fixed. If additional feeder capacity is needed suddenly, the risers constitute a difficult bottleneck. Only so much energy is going to be able to get through. New risers are very expensive to install and present many practical problems.

Extra feeder capacity can be incorporated without running extra copper wire. The design can be drawn to select raceways that are larger than needed at the time or installation. Then, when the extra feeder capacity is required, the raceway is already in place and it is simply a matter of pulling more copper.

Combining extra raceway capacity and initial installation of thicker than required copper wire is the way that future needs are usually prepared for. In addition, the feeder distribution centers, switchboards, and branch circuit panelboards can be made larger, so that the demands of the coming years can be satisfied.

STREET WIRING

Each of us as an individual consumer is interested in the way that electrical energy is conveyed along the urban thoroughfares from generating plant to residence. The subject of electrical wiring would not be complete without an explanation of the methods, routes, equipment and appurtenances by which city dwellers receive their electrical power.

POLES

The wiring between the utility company and the rack of insulators below the consumer's weatherhead is carried on the support arms of the familiar electric pole. (In the most modern systems, the wires are conveyed in pipes laid underground.) The pole or duct line transmits the energy through copper conductors.

At first glance a pole looks like a simple object but there is a lot more there than meets the eye. Poles come in three main types: wood, steel, and concrete. Wood and steel poles are rated to have a life of 20 to 30 years. A concrete pole is virtually immortal, and it is the most expensive of the three.

The length of the pole to be used depends upon the function(s) it will perform. About one-sixth of the pole goes in the ground. Along highways poles usually rise above the ground 18 to 21 feet. The height of the pole is sometimes related to the number of crossarms it will be asked to carry. The accompanying charts give rule-of-thumb heights (Fig. 18-1). The local municipal ordinances regulate the poles.

Poles are usually arranged in straight lines and spaced about 125 feet apart. The intervals vary between a maximum of 150 feet and a minimum of 100 feet.

The height and the spacing is dependent on the terrain. In hilly areas short poles are used on the hilltops and longer poles in the valleys. The idea is to keep the wires running as straight as possible.

The span between poles on curves and around corners is given in the table. The angle of turn and the pull in feet with respect to the span have all been calculated and installed many times.

Class	1	2	3	4	5	6	7	8	9	10
Min top circumference in..	27	25	23	21	19	17	15	18	15	12
Length of pole, ft.	Min circumference 6 ft. from butt, in.									
Northern White Cedar Poles										
16	....	....	....	....	26.0	24.0	22.0	No butt requirement		
18	....	....	32.5	30.0	28.0	25.5	23.5			
20	39.5	37.0	34.0	31.5	29.0	27.0	25.0			
22	41.0	38.5	36.0	33.0	30.5	28.0	26.0			
25	43.5	41.0	38.0	35.5	32.5	30.0	28.0			
30	47.5	44.5	41.5	38.5	35.5	33.0	30.5			
35	50.5	47.5	44.0	41.0	38.0	35.0	32.5			
40	53.5	50.0	46.5	43.5	40.0	37.0				
45	56.0	52.5	49.0	45.5	42.0					
50	58.5	55.0	51.5	47.5	44.0					
55	61.0	57.5	53.5	49.5	46.0					
60	63.5	59.5	55.5	51.5						
Western Red Cedar Poles										
16	....	....	....	....	23.0	21.5	19.5	No butt requirement		
18	....	....	28.5	26.5	24.5	22.5	21.0			
20	34.5	32.0	30.0	28.0	25.5	23.5	22.0			
22	36.0	33.5	31.5	29.0	27.0	25.0	23.0			
25	48.0	35.5	33.0	20.5	28.5	26.0	24.5			
30	41.0	38.5	35.5	33.0	30.5	28.5	26.5			
35	43.5	41.0	38.0	35.5	32.5	30.5	28.0			
40	46.0	43.5	40.5	37.5	34.5	32.0				
45	48.5	45.5	42.5	39.5	36.5					
50	50.5	47.5	44.5	41.0	38.0					
55	52.5	49.5	46.0	42.5	39.5					
60	54.5	52.5	47.5	44.0						
65	56.0	51.0	49.0	45.5						
70	57.5	54.0	50.5	47.0						
75	59.5	55.5	52.0	48.5						
80	61.0	57.0	53.5	49.5						
85	62.5	58.5	54.5							
90	63.5	60.0	56.0							

Fig. 18-1. Utility poles ordinarily rise to heights of 16 to 25 feet. In exceptional cases, though, 90-foot long poles may be used.

Pin holes				Center Bolt Hole, in.	Brace Length, in.	Size and Length
Spacings, in.			Size, in.			
Center	Sides	Ends				
Electric Light Arms						$3\frac{1}{4} \times 4\frac{1}{4}$
28	....	4	1 17/32	⅝	25	3 ft 2 pin
16	12	4	1 17/32	⅝	28	4 ft 4 pin
18	17	4	1 17/32	⅝	28	5 ft 4 pin
22	21	4	1 17/32	⅝	32	6 ft 4 pin
16	12	4	1 17/32	⅝	32	6 ft 6 pin
18	17½	4	1 17/32	⅝	32	8 ft 6 pin
16	12	4	1 17/32	⅝	32	8 ft 8 pin
16	9¾	4	1 17/32	⅝	32	8½ ft 10 pin
17½	15¾	4	1 17/32	⅝	42	10 ft 8 pin
16	12	4		⅝	42	10 ft 10 pin
16	9⅝	3⅞	1 17/32	⅝	42	10 ft 12 pin
R.S.A.Arms						$3 \times 4\frac{1}{4}$
20	22	4	9/16	11/16	...	6 ft 4 pin
19	17¼	4	9/16	11/16	...	8 ft 6 pin
19	15½	4	9/16	11/16	...	10 ft 8 pin
16	12⅜	2½	9/16	11/16	...	10 ft 10 pin
Western Union Arms						$3 \times 4\frac{1}{4}$
20	11½	3	9/16	21/32	...	6 ft 6 pin
21	11½	3	9/16	21/32	...	8 ft 8 pin
22	11½	3	9/16	21/32	...	10 ft 10 pin
Pony Telephone Arms						$2\frac{3}{4} \times 3\frac{3}{4}$
17	....	3½	1 9/32	⅝	...	24 in. 2 pin
23	...	3½	1 9/32	⅝	...	30 in. 2 pin
29	...	3½	1 9/32	⅝	25	36 in. 2 pin
16	9½	3½	1 9/32	⅝	28	42 in. 4 pin
16	9¾	3½	1 9/32	⅝	28	62 in. 6 pin
16	9¾	3¾	1 9/32	⅝	28	82 in. 8 pin
16	9¾	4	1 9/32	⅝	28	102 in. 10 pin
16	9⅝	3⅞	1 9/32	⅝	28	120 in. 12 pin
N.E.L.A. Arms						$3\frac{1}{2} \times 4\frac{1}{2}$
30	....	4	1 17/32	11/16	28	3 ft 2 in. 2 pin
30	14½	4	1 17/32	11/16	38	5 ft 7 in. 4 pin
30	14½	4	1 17/32	11/16	38	8 ft 6 pin
30	12	4	1 17/32	11/16	38	9 ft 2 in. 8 pin

Fig. 18-2. This list of specifications for the manufacture of standard wooden crossarms attests to the product's importance to the electrical industry. (Courtesy of General Electric Supply Co.) Continued on next page.

						N.E.L.A. (Light) Arms	$3\frac{1}{4} \times 4\frac{1}{4}$
30	....	4	1 17/32	11/16	28		3 ft 2 in. 2 pin
30	14½	4	1 17/32	11/16	38		5 ft 7 in. 4 pin
30	14½	4	1 17/32	11/16	38		8 ft 6 pin
30	12	4	1 17/32	11/16	38		9 ft 2 in. 8 pin
			New England Arms				$3\frac{1}{4} \times 4\frac{1}{4}$
30	....	3	1 17/32	11/16	33		3 ft 2 pin
30	13½	4½	1 17/32	11/16	36		5 ft 6 in. 4 pin
30	13½	4½	1 17/32	11/16	36		7 ft 9 in. 6 pin
30	13½	4½	1 17/32	11/16	36		10 ft 8 pin
			New England Power Arms				$3\frac{3}{4} \times 4\frac{3}{4}$
30	....	3	1 17/32	11/16	33		3 ft 2 pin
30	13½	4½	1 17/32	11/16	36		5 ft 6 in. 4 pin
30	13½	4½	1 17/32	11/16	36		7 ft 9 in. 6 pin
30	13½	4½	1 17/32	11/16	36		10 ft 8 pin
			Pacific Arms				$3\frac{1}{4} \times 4\frac{1}{4}$
28	...	4	1 17/32	⅝	32		3 ft 2 pin
28		4	1 17/32	⅝	32		5 ft 4 pin
28		4	1 17/32	⅝	32		˙ 7 ft 6 pin
28		4	1 17/32	⅝	42		9 ft 8 pin
28		4	1 17/32	⅝	42		˙ 11 ft 10 pin

Fig. 18-2 cont.

CROSSARMS

At the top of most poles is the familiar crossarm. The crossarm is not just a strip of wood nailed near the top of the pole. It is a carefully engineered, treated, seasoned, precisely painted piece of Douglas fir or pine wood.

Surprisingly, the crossarm has not been standardized and is subject to the design of the engineer who has a particular use for the arm. When crossarms are ordered from the manufacturer, either an engineering sketch is provided by the client or one is drawn up by the manufacturer.

The General Electric Supply Co. drew up the standard wooden crossarm table of Fig. 18-2. A lot of considerations go into the design of a crossarm.

First of all, there has to be a space in the center of the arm for a lineman to climb into and through. The National Electric Light Association believes 30 inches is a safe climbing space. No wires are allowed in that space.

Crossarms have a center hole and a bolt passes through this and the pole. On either side of the center hole is a brace hole for a bolt which fastens the crossarm to the brace.

The center hole and brace holes are horizontal. In addition, the crossarm has vertical holes. These are the holes by which the insulators are bolted. The wiring is suspended across the insulators (Fig. 18-3).

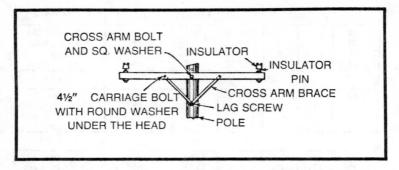

Fig. 18-3. Braces and insulators are invariable features of utility pole crossarms.

If the service is 2-wire, single-phase, only two insulators are needed on either end of the crossarm. To add additional lines for a 3-wire or 4-wire service, two more top-to-bottom holes can be drilled and insulators installed. The new holes are 14½ inches closer to the center of the crossarm.

The single crossarm is used where runs are straight and not across obstructed terrain. If the load becomes heavy or turns are made in the run, a double crossarm can be used. Typical locations for double crossarms are at street corners and railroad crossings.

Two single crossarms can be separated by wooden blocks, spacing bolts, or spacing nipples. Avoiding the use of double arms as much as possible is desirable. Although it makes the installation much stronger, the double arm restricts climbing space and is more difficult to work with. Since the lineman's duties will take place 20 feet in the air, it is safest to make his work as easy as possible.

LOCATION OF CIRCUITS

The through lines are installed on the uppermost crossarms. The local wires that are tapped off to the user are located on the lower crossarms.

For a particular 2-wire or 3-wire ciricut, the wires are kept near each other on adjacent insulators. This is important because, besides the ease of handling, the inductance of the wire run gets larger as the distance of the run increases. If the wires of a circuit are far apart, the inductance becomes large and a severe voltage drop takes place. When the wires are close together, the two inductances opposite in phase tend to cancel out each other. This reduces the voltage drop and saves a lot of energy.

As the wires run from pole to pole, they should be installed on the same insulator positions. If they are not, much confusion takes place. The neutral wire should always be mounted between the two phase lines in a 3-wire circuit.

The racks of insulators that carry the low voltage secondary to the user are mounted on the side of the pole and not on the cross-arms. The illustrations show a few of the different types of installations (Fig. 18-4).

The distribution transformers, or as they are called, the pots, are also mounted on the sides of the poles. A convenient position is between the top crossarm and the bottom crossarm.

The pot is a step down transformer. The top crossarm carries the through lines which are also the primary voltage. This is high voltage. The high voltage can be attached to the primary of the transformer and a primary fuse installed in the primary line.

The secondary voltage of the pot, which could be a 120/240-volt, 3-wire type, is then run to the bottom crossarm. From the bottom crossarm to secondary insulators on the side of the pole, the 3-wire system is run.

Other ways to install the pot are below the bottom crossarm and on the top crossarm. Each mounting location is dependent on the features of the particular installation.

DISTRIBUTION SYSTEMS

Electricity in common use has voltages that range from 110 to 275,000. There are a couple of dozen different quantities of voltages crackling along the highways and across the countryside.

The lower end voltages are carried by poles or underground ducts; the higher kinds of voltage travel over long distances on high aerial wires or towers.

Practically all of the voltage is AC, although there has been interest in long distance DC. Some foreign nations have started to use long distance DC. It will be a long time before the U.S. goes into it. There are some advantages, but discussing them is beyond the scope of this book.

You can assume that most power voltages that are transmitted in the U.S. are AC. In addition, the AC is 60 cycle, or 60 hertz as it is now starting to be called.

There are some 25-hertz installations in railroad applications, steel mills, and cement mills. The 25-hertz frequency has fewer

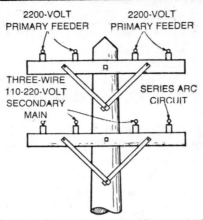

2200-VOLT PRIMARY FEEDER   2200-VOLT PRIMARY FEEDER

THREE-WIRE 110-220-VOLT SECONDARY MAIN

SERIES ARC CIRCUIT

LOCATION OF CIRCUITS ON TWO-ARM POLE

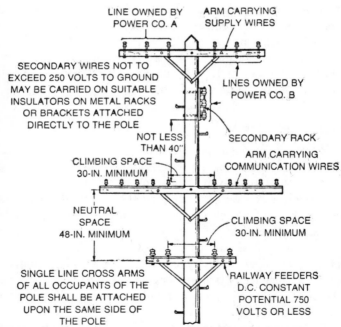

LINE OWNED BY POWER CO. A

ARM CARRYING SUPPLY WIRES

SECONDARY WIRES NOT TO EXCEED 250 VOLTS TO GROUND MAY BE CARRIED ON SUITABLE INSULATORS ON METAL RACKS OR BRACKETS ATTACHED DIRECTLY TO THE POLE

LINES OWNED BY POWER CO. B

NOT LESS THAN 40"

SECONDARY RACK

ARM CARRYING COMMUNICATION WIRES

CLIMBING SPACE 30-IN. MINIMUM

NEUTRAL SPACE 48-IN. MINIMUM

CLIMBING SPACE 30-IN. MINIMUM

SINGLE LINE CROSS ARMS OF ALL OCCUPANTS OF THE POLE SHALL BE ATTACHED UPON THE SAME SIDE OF THE POLE

RAILWAY FEEDERS D.C. CONSTANT POTENTIAL 750 VOLTS OR LESS

CONSTRUCTION OF JOINTLY USED WOOD-POLE LINES CARRYING SUPPLY CIRCUITS SIGNAL CIRCUITS, SHOWING ARRANGEMENT OF ATTACHMENTS AND CLEARANCE (HUBBARD & CO.)

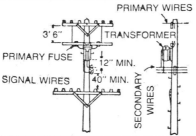

PRIMARY WIRES

3' 6"   TRANSFORMER

PRIMARY FUSE

12" MIN.

SIGNAL WIRES   40" MIN.

SECONDARY WIRES

SINGLE PHASE TRANSFORMER MOUNTED
IN SPACE BETWEEN PRIMARY WIRES AND
SECONDARY RACK. (LINE MATERIAL INDUSTIRES)

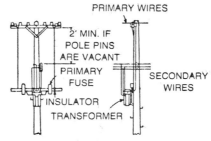

PRIMARY WIRES

2' MIN. IF
POLE PINS
ARE VACANT

PRIMARY
FUSE

SECONDARY
WIRES

INSULATOR

TRANSFORMER

SINGLE PHASE TRANSFORMER MOUNTED
BELOW SECONDARY RACK. (LINE MATERIAL
INDUSTRIES)

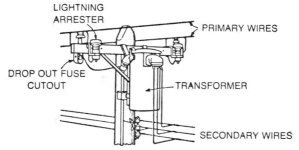

LIGHTNING
ARRESTER

PRIMARY WIRES

DROP OUT FUSE
CUTOUT

TRANSFORMER

SECONDARY WIRES

SINGLE PHASE TRANSFORMER MOUNTED
ON SAME CROSS ARM WITH PRIMARY
SUPPLY WIRES. (GENERAL ELECTRIC CO.)

Fig. 18-4. The arrangement of crossarms, braces, insulators, arresters and pots
depends upon the situation.

inductive effects and can drive large slow-speed motors such as those used in the heavy industries just mentioned. This frequency, though, when used in lighting causes flickering. The filament cools down between cycles and the lights waver, which is hard on workers' eyes.

The standard frequency is 60 hertz. Any 50-cycle installations that are left over from years ago are being changed over to 60 hertz.

If a higher frequency is needed for a special application such as high speed motors or fluorescent lamps, a frequency converter can be used to increase the number of hertz.

The 60-hertz systems are: the common 120/240-volt, single-phase; 120/208-volt, three-phase; 240-volt, three-phase; 480-volt, three-phase; and 277/480-volt, three-phase.

There are a number of industrial-type installations, such as: the 2400-volt, three-phase; 2400/4160-volt, three-phase; 4800-volt, three-phase; 7200-volt, three-phase; and 7.2/13.8-kilovolt, three-phase. The higher voltages are fed to the industrial plant, which has its own substation to step down the voltages for the particular needs.

GROUNDING THE SYSTEMS

The electrical transmission system up and down the streets is grounded. Every individual user is also grounded. Grounding gives all the systems the same zero voltage point and ties them together.

Ideally every part in every system that can be contacted easily should be grounded. Only the live phase lines are not grounded. The live lines are supposed to be satisfactorily insulated from ground. The Code requires all types of grounding and Article 250 goes into great detail. When all systems are correctly grounded, safety is at the maximum.

Three common hazards are lightning strikes, breakdown in the insulation between primary and secondary windings in transformers, and accidental contact between high voltage wires and low voltage wires. All three hazards can produce fire, explosion, and death.

When the systems are correctly grounded, lightning finds a path to ground and is discharged harmlessly. Transformer short circuits will blow out the primary fuse and open the circuit. Short circuits between wires carrying large differences in voltage open up a circuit breaker, fuse or ground fault circuit interrupter.

The primary circuit on the pole is grounded at the transformer. The secondary is grounded at the customer's water pipe.

If the insulation in the pot between the primary and secondary should break down, the two different voltage levels, the 2300 volts and the 240 volts, will make contact.

If there were no ground, the 2300 volts would wreak havoc in the secondary. The secondary is only insulated for 600 volts and the 2300 would burst through the insulation and cause all manner of damage. If anyone were touching a fixture or appliance at the time, death would be the result.

GROUND RODS

The ground rod used in street wiring is usually a steel rod with a coating of pure copper molten welded to the steel cone. The copper coating makes a good electrical connection to ground and resists corrosion from the weather and earth (Fig. 18-5).

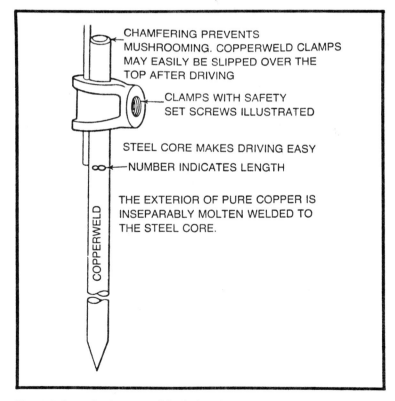

CHAMFERING PREVENTS MUSHROOMING. COPPERWELD CLAMPS MAY EASILY BE SLIPPED OVER THE TOP AFTER DRIVING

CLAMPS WITH SAFETY SET SCREWS ILLUSTRATED

STEEL CORE MAKES DRIVING EASY

NUMBER INDICATES LENGTH

THE EXTERIOR OF PURE COPPER IS INSEPARABLY MOLTEN WELDED TO THE STEEL CORE.

COPPERWELD

Fig. 18-5. Ground rods are carefully designed and carefully installed to confer the greatest safety.

The ground rod makes a low resistance contact with the earth. The copper water pipe system is also making a low resistance contact with earth and all are locked in together at zero volts.

The rod is pointed at the bottom and is marked along the shaft with a number indicating its length. An important component of the rod is a clamp which can be solidly tightened against the rod with a setscrew. The ground wire from the pole also fits into the clamp. The ground wire is pressure connected to the clamp and ground rod in this way.

The end of the rod is chamfered. This means it is beveled so the end will not spread out if the rod must be driven into rocky soil. The clamp can be slipped on after the rod is driven without fear that the rod end will have been expanded by the maul.

GROUND WIRES

The ground wire is not permitted to be thinner than #8 copper and some companies will not use anything thinner than #4 copper. Bare wire is satisfactory for the run down the pole to the rod. The wire is secured to the pole with cleats or straps.

There is a ground for each transformer group. A ground must be attached every 500 feet if there is a neutral wire in the supply.

Under certain soil conditions a shock can be received by anyone standing near the ground wire. As a result, the wire is encased by wooden molding for a distance of at least 8 feet from the surface. In fact, the entire length of ground wire should be in wooden molding for the protection of the lineman.

GROUND RESISTANCE

The Code would like, but does not demand, a ground resistance measurement every time a ground is installed. The ground resistance is the number of ohms between the ground rod and the earth. The smaller the resistance the better.

The closer the resistance is to zero ohms, the closer the ground rod voltage is to zero volts. The Code would like a maximum resistance of 3 ohms for water pipe grounds and a maximum of 25 ohms for buried rod grounds.

A 25-ohm ground is not a good ground. It is passable. The ground resistance is dependent on the nature of the soil. Pure sand has a high resistance and is a terrible ground. Moisture laden, rich soil is a good ground because its resistance is low.

There are a number of ways to reduce a high ground resistance. None of the methods is surefire, but little by little you can lower the ground resistance.

First of all, increase the diameter of the ground rod. This provides more surface area to contact the earth and lowers the resistance.

Second, use longer ground rods and try a second, third, or fourth ground rod, all wired parallel. Each ground rod places more surface area against the earth.

Third, install the rod at a damp site; for example, under a rainwater runoff where the ground will remain soaked most of the time.

Fourth, treat the earth with a chemical that will infuse a permanent low resistance factor in the earth. Chemicals such as magnesium sulfate, copper sulfate, and common rock salt are all good conductors. When mixed with the soil around the ground rod, they lower the resistance between the pipe and the earth.

When the ground resistance is high, fuses and circuit breakers do not open during faults. A ground circuit return to earth with a high resistance keeps the fault current in the installation: conduit, boxes and other metal parts will become live. The importance of a good low resistance ground cannot be emphasized enough.

UNDERGROUND WIRING

Underground wiring is desirable. Many companies are installing and converting as much overhead wiring to underground as possible.

There are three ways of laying underground wiring in the street. One way takes advantage of existing tunnels to lay conduit inside. On occasion a tunnel is dug specifically for the wiring.

By the second way, trenches are dug and conduit is buried in the trench. By the third way, cable is buried in a trench.

The conduit in a trench is called a duct line. The conduit in the duct can be made from any of a number of materials: fiber, tile, iron, asbestos, PVC, PE, styrene, and concrete.

The PVC (polyvinylchloride), the PE (polyethylene), and the styrene are all popular types of duct because they are inexpensive, easy to install, and available in 30-foot lengths (The long lengths represent savings in the purchase and installation of couplings.)

Iron is too expensive for most projects, but it does give the cables good protection from heavy machinery. Iron is often used in parkway lighting-circuit runs in fill earth.

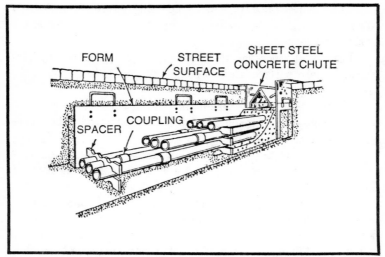

Fig. 18-6. Many of the electric lines that were once strung from poles are now run underground. The duct lines are provided with access at convenient intervals.

An asbestos product called Transite is a combination of asbestos and cement. It is strong, light, corrosion resistant, incombustible, noninductive and has good heat conductivity. It can be used above or beneath the soil.

Each type of duct line is best for a particular application. When a duct line is designed, all the characteristics are taken into consideration and a conduit composed of the most appropriate material is chosen. The characteristics and composition of conduit can be obtained from the manufacturers and suppliers.

INSTALLING A DUCT LINE

A duct line is a run of conduit below the earth surface. A trench is excavated so that the top of the duct line is at least 2 to 3 feet below the surface. The duct line is graded so that it pitches forward at a slight angle. About 1 foot in 100 feet is sufficient. This ensures a good liquid drainage.

The duct line should be in contact with soil at least along one side. The soil will serve to convey away any excess heat that might develop as current passes through the conductors.

Usually duct lines have a number of conduit runs, all clustered together. There should be at least 1 to 3 inches of earth or concrete between the different conduits. In case a short circuit should occur in one conduit, flames, arcing, and heat will not be able to affect the adjacent ducts and spread the troubles. If the line is of high voltage, at least 3 inches of separation should be maintained. A 1-inch separation is satisfactory for lower voltages.

The duct line runs of conduit are held in place by plastic spacers during construction. Sometimes the spacers are removed after the construction, but at other times the spacers are left in place and become a permanent part of the installation (Fig. 18-6).

The firmness of the soil is a consideration. When the soil is not firm, forms must be used to hold the soil. (Sandy soil is especially yielding.) When the soil is firm, the soil itself acts as a form.

Concrete is poured below and around most types of conduit. The exception is Transite. Concrete is not needed for Transite unless the soil is not firm. For the trench details, various construction manuals should be consulted.

MANHOLES

A manhole is a specially excavated and sized room in the ground. All the ducts enter the manhole and the wiring is accessible there. The manhole is designed to allow the lineman to perform most of the wiring work necessary during service and maintenance.

Manholes are constructed of many sizes and shapes to fit particular local conditions. After all is said and done, the style and size of the classical manhole in Fig. 18-7 is the best for average conditions.

The largest of the common-purpose manholes is about 5' × 7'. The smallest is about 3' × 4'. If a transformer is to be located inside a manhole, the size of the manhole should be increased accordingly.

Manholes are made of brick, concrete, or a combination of both. Concrete is the cheapest, but the best manholes have concrete bottoms, brick sides, and concrete tops. Manholes have been so standardized that they are precast and shipped by the manufacturer to the construction site.

The shape of a manhole is normally round rather than square or rectangular. This is to allow the wiring and duct work to run smoothly around the manhole without encountering sharp bends and corners. This makes it easier for the lineman to work with large copper wires.

MANHOLE COVERS

The manhole head is cast from iron or steel, with steel being the preferred type but more expensive. Manhole covers are round so

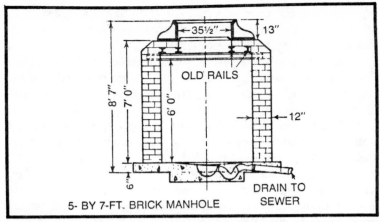

Fig. 18-7. A good general-purpose style of manhole has a concrete bottom, brick sides and concrete top.

they cannot fall into the hole. Square and rectangular covers caused damage to equipment and harm to workers in years past.

The manhole cover prevents rain from falling into the room but is not watertight. The covers are not to be fastened down in any event. In case of accumulated gas and subsequent detonation, a manhole cover is designed to blow off and vent the force of the explosion. In addition there are ventilation holes in the cover, so gases can escape freely and not build up. The ventilation holes are slots and can take up 50 percent of the cover area.

Dirt and water do enter the rooms and can cause an uncomfortable, infested environment. However, manholes can be cleaned out and they present little or no safety hazard. Sealing up the manhole, on the other hand, could produce a situation in which the unit would become a virtual bomb.

Since the manhole is expected to accumulate water and dirt, a sewer connection should be installed at the bottom of every manhole. The sewer runoff can be an elaborate arrangement with a trap and strainer or it can simply be a drain hole in the bottom of the room. When a drain hole is used, a pocket should be dug out beneath the room and filled with broken rock. This is a form of dry well.

INSTALLING CABLE

Cable in the manhole is usually heavy. Pulling the cable and getting it connected is hard work. The cable should be bent as little as possible.

The points where the conduit enters the manhole should be chamfered so the cable is deformed as little as possible. The cable is pulled with the aid of pull-in wires. Great care is taken to make sure the duct lines are clear and clean. The cable is greased, then pulled by manpower, winches, trucks, or even tractors.

Pulling a number of cables into one duct line would seem to be an easy task. Removing them when necessary would also seem to be easy. However, once a cable is nestled within a group of cables, it is really immovable. Over a long period of time, dirt and grit binds the cables together and, if you try to remove one or more, you strip the sheathing from all the adjoining cables and cause short circuits and other hazards.

As a result, the installation is designed so that each cable can lie in an exclusive duct. Then when a cable is removed, the single cable could be ruined, but it would be the only casualty.

The layout of the cables should be carefully planned so that there is an absolute minimum of crossover. In addition, the cable is best run as straight as possible with the least amount of bending. Each bend increases the length of the cable, which in turn increases losses, adds to the ambient temperature, and enhances magnetic effects.

Racks are installed in manholes to support the cable as it passes from one duct line to another. The rack is a steel frame with porcelain insulator arms.

Index

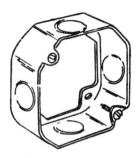

Index

15542